# GIS and Housing

*GIS and Housing: Principles and Practices* discusses one of the challenges that has not been addressed by Geographic Information Science thus far: how can we use GIS to deal with the complex issues underlying the housing crisis? This book provides GIS technicians and analysts with an overview of US housing challenges and examples of how to effectively integrate spatial thinking to address housing policy questions, while simultaneously introducing housing policy analysts to advanced GIS concepts and techniques to create livable neighborhoods that include housing alternatives beyond the single family. Through numerous examples, the authors advocate for a collaborative approach that encourages professionals, policymakers, and analysts, across different ideological and political perspectives, to confront the multifaceted housing crisis.

Features:

- Examines the historical aspects of housing provision, societal attitudes, demographic shifts, and government policies.
- Bridges the gaps between housing professionals and GIS experts, facilitating an interdisciplinary approach to address the housing crisis.
- Explores different challenges that are facing urban, suburban, and rural neighborhoods in different US regions.
- Provides professionals with the necessary tools for informed decision-making.
- Proposes solutions that leverage the integrative capacity of GIS to address established housing issues.
- Advocates for denser housing alternatives to address issues of affordability, supply shortages, and homelessness.

This book is intended for graduate students and professionals in housing, community development, urban planning, architecture, and GIS, and anyone curious about learning more about the American housing crisis.

## GIS in Action

*Series Editor*

*GIS and Housing: Principles and Practices*
Laxmi Ramasubramanian, Jochen Albrecht, and Deborah Rojas De Leon

For more information about the series, please visit: https://www.routledge.com/GIS-in-Action/book-series/CRCGISINACTI

# GIS and Housing
## Principles and Practices

Laxmi Ramasubramanian, Jochen Albrecht,
and Deborah Rojas De Leon

CRC Press
Taylor & Francis Group
Boca Raton London New York

CRC Press is an imprint of the
Taylor & Francis Group, an **informa** business

Designed cover image: © The Authors

First edition published 2024
by CRC Press
2385 NW Executive Center Drive, Suite 320, Boca Raton FL 33431

and by CRC Press
4 Park Square, Milton Park, Abingdon, Oxon, OX14 4RN

*CRC Press is an imprint of Taylor & Francis Group, LLC*



*Library of Congress Cataloging-in-Publication Data*
Names: Ramasubramanian, Laxmi, author. | Albrecht, Jochen, author. |
De Leon, Deborah Rojas, author.
Title: GIS and Housing : principles and practices / Laxmi Ramasubramanian, Jochen Albrecht, and Deborah Rojas De Leon.
Description: 1 Edition. | Boca Raton, FL : CRC Press, [2024] | Series: GIS in action | Includes bibliographical references and index. | Summary: "This book discusses how GIS enhances our understanding of complex housing issues in the United States. It provides an overview of US housing issues and examples of how to effectively integrate spatial thinking to address housing policy questions, while simultaneously introducing housing policy analysts to advanced GIS concepts and techniques to create livable neighborhoods that include housing alternatives beyond the single family. Through numerous examples, the authors advocate for a collaborative approach that encourages professionals, policymakers, and analysts, across different ideological and political perspectives, to confront the multifaceted housing crisis"– Provided by publisher.
Identifiers: LCCN 2023036808 (print) | LCCN 2023036809 (ebook) |
ISBN 9780367197339 (hardback) | ISBN 9781032676562 (paperback) | ISBN 9780429243066 (ebook)
Subjects: LCSH: Housing. | Community development. | City planning. |
Geographic information systems.
Classification: LCC HD7287. R326 2024 (print) | LCC HD7287 (ebook) |
DDC 363.5–dc23/eng/20230911
LC record available at https://lccn.loc.gov/2023036808
LC ebook record available at https://lccn.loc.gov/2023036809

ISBN: 978-0-367-19733-9 (hbk)
ISBN: 978-1-032-67656-2 (pbk)
ISBN: 978-0-429-24306-6 (ebk)

DOI: 10.1201/9780429243066

Typeset in Times
by codeMantra

# Contents

# Preface

The three authors of this book are professionals in the realms of architecture, geospatial analysis, and urban planning. The publisher invited us to propose a book about GIS applications and left it up to us to choose the application area. In considering our different areas of expertise, we began to hone in on housing. Preliminary investigations revealed a glaring gap in recent GIS applications literature: we could not find a book-length treatment of GIS applications for housing. The reasons for the gap became gradually apparent as we conceptualized and framed this book. First, since housing production is largely left to the market, much of the data and analysis about that sector tend to be proprietary and not readily accessible to the public. Second, housing policy analysis is a highly specialized sub-genre that typically focuses on affordable housing, assessing the impacts of government policies and programs designed to improve housing affordability or the assessment of government policies that attempt to remove structural or institutional barriers to housing affordability. Lastly, the financing of housing production using federal and state-level data that dominate policy conversations, subsuming design and planning considerations that rely on local and sub-regional data.

The study of housing as a GIS application area has many opportunities and challenges. Undoubtedly, the study of housing is central to other fields such as economic development, transportation, and crime/public safety. As the field of GIS grew and matured in the 1980s and 1990s, GIS specialists, particularly those scholars interested in GIS applications, were actively involved in shaping GIS policies to increase access to spatially referenced data. For example, GIS specialists analyzed home lending data made publicly available through the Housing Mortgage Disclosure Act. These analyses and insights made discriminatory lending practices visible to the general public and to lawmakers. However, in the past two decades, GIS applications in housing appear to have not received much attention.

This book speaks to a new generation of GIS users and specialists who have grown up in a world where the early challenges of spatial data access have largely been resolved. In addition to Census data that is publicly available, a range of datasets generated for different purposes can be accessed and linked using a common spatial identifier. The advent and democratization of geospatial technologies provide us for the first time with the tools to deal with housing in the context of larger societal shifts. Current shortcomings in the provision of adequate shelter for everyone cannot be addressed without seeing its embeddedness in questions of demographic changes (immigration, aging societies, and homelessness), climate change, or the impact that information technology has on labor markets, transportation, and individualized services. Yet, data alone is not sufficient to prompt interesting and intelligent queries – a deep understanding of the phenomena being investigated is also necessary. Otherwise, GIS specialists can develop solutions to non-existent problems or worse, arrive at erroneous conclusions because they do not fully understand the social phenomena under scrutiny. GIS applications research requires that GIS specialists understand the world as it is, not the world that is accessible through the

GIS interfaces. The real world is far more complex than the comfortable vector GIS landscape that comprises points, lines, and polygons.

This book engages housing researchers, alerting them to how the GIS technology and data landscape have changed and encouraging them to go beyond simple mapping and overlays of phenomena. Asking, "where are all the public housing properties in the city located?", is a useful starting point, but GIS in the 2020s can be tasked to do much more. Complex queries and new lines of inquiry require that domain specialists (in this case, housing experts) and GIS specialists work in partnership to resolve pressing social problems such as homelessness.

Our diverse perspectives invite readers from various fields to delve into these pages, exploring the important and often missed interconnections between housing and broader societal shifts that impact people at the neighborhood level. We have written this book using accessible and jargon-free language with a wide range of examples from big- and medium-sized cities as well as small towns and rural areas. It is our fervent hope that elected officials and decision makers interested in pragmatic problem solving will read this book. We encourage readers to understand our perspectives – GIS tools and spatial thinking allow end users to swiftly move between and across spatial scales to identify actionable policy levers appropriate to solve the problem at hand. Private residential housing production and management is largely a local matter in the United States – and therefore conversations about densification should occur at the local level. National or state mandates about densification notwithstanding, the preferred housing type in America is a spacious, single family home. We encourage policymakers to focus on encouraging a diversity of housing alternatives, emphasizing new designs and new ownership models. We also encourage decision makers to use the integrative potential of geospatial technologies to explore the challenges that are coming toward us rapidly – demographic changes worldwide and the global climate change impact the housing situation in the United States and no enclave can be immune to these effects. In other words, housing insecurity and homelessness will continue to increase and it has to be confronted at the local/regional scale in order to have quick and meaningful impacts.

GIS and housing specialists are focused on numbers; this is unsurprising because quantification is essential to make a case for large investments of public dollars. In this book, we encourage architects, landscape architects, and urban designers to engage with housing and GIS specialists in order to co-create innovative design alternatives, for example, by investigating new living options for the 55+ and over residential market becomes critical as our national demographic trends shift.

We end this book with a note to students – future architects, planners, engineers, GIS specialists, and aspiring elected officials. We wrote this book with you in mind. As educators, each of us has worked with hundreds of individual students and we poured our collective knowledge, experience, and expertise into the pages of this book. We filled our pages with dozens of examples from all parts of the United States; so, you can find the context that relates to your circumstance and location. We have provided cross-referencing within the book as well as literally hundreds of references for further reading. We have developed a companion website (gisandhousing.com), where we plan to provide updates, errata, and further examples. In the long run, we plan to write a follow-up volume of GIS exercises that go beyond the limited amount of how-to's that we could include in this volume. Please use our website to engage with us as we strive to keep the contents of this book current.

Housing is deceptively simple as it is complex. Consider "poor doors", "dorms without windows", restrictive housing covenants, or the power that homeowners' associations wield and it becomes quickly apparent that our own values shape and influence housing policy as well as our solutions to serve the most vulnerable among us. If we are going to tackle the housing crisis, developers and real estate professionals have to work in partnership with stakeholders in all levels of government, and the nonprofit sector. We encourage a geospatial perspective as a lens to tackle the housing crisis. Our diverse perspectives invite readers from various fields to delve into these pages, exploring the important and often missed interconnections between housing and broader societal shifts that impact people at the neighborhood level. Our aim is to empower readers to apply a geospatial framework to confront the housing crisis. We envision a future where housing becomes a right accessible to all, fostering a more just and inclusive world.

# Foreword

As a planner, an advocate, an administrator, and a former political appointee I've stepped out in front of scores of boards, hearings, working groups, and meetings of many types to attempt to secure approvals in the service of getting more housing built. Often there is quick agreement amongst everyone that a home is essential to provide stability and safety, and that the barriers to housing, particularly the increasing cost to rent or own, need urgent attention and intervention. Together, we'll exchange analogous statements that making housing more accessible will strengthen the health, fiscal, and societal bonds of a community. But despite that common ground, it is not long before too many of these discussions can devolve into perplexing contentiousness. In these exchanges about whether housing should be allowed in the proposed location, designed as suggested, and serving the mix of people we're proposing to serve, it's critical that we leverage our most reliable and persuasive tools if there is any chance these proposals will be embraced.

The high bar is because the subject of housing – yours or someone else's – can be incredibly personal, the arguments as subjective and varied as the gamut of those, with their individualized experience and values, that present them. The debates that play out occur in exchanges equally driven by facts as they are by feeling. For many it's not simply a matter of public policy or rational planning, but a decision that represents the most consequential thing standing between themselves and protecting their prosperity. The potential of new housing can be seen as both an opportunity and a risk – this perspective sometimes shifts from support to opposition and vice versa when talking about different geographies where a proposal may be considered. In those moments, how the information is presented, how responsive it is to address broader concerns, and the credibility of that information can make or break a proposal.

The act of holding the discussion is not the problem, it remains part of the solution. It should occur through direct participation and elected representatives evaluating the complexities of broad regulatory and tax reforms or more discrete discretionary actions that can potentially unlock barriers to growing the housing supply. And depending on the scope of the changes proposed, conversations go beyond a decision about the buildings themselves, but instead drill into questions about local infrastructure, public transit, parks, roads, sustainability, and school seats which either through mandated environmental reviews or voluntarily offered research, bring to the surface some reasonable, and difficult, questions that need to be considered alongside the need for housing.

These forums are at their most effective when participants are supplied with data that is vetted and factual and not primarily driven by anecdotes or assumptions. This is not to say there isn't a role for qualitative techniques and descriptive input. Non-numerical information can be invaluable and needs to be integrated to fully grapple with the complex questions being considered. But it's the mixed methods approach, with dynamic analysis at its foundation, that can allow for a faster, fuller charting of where your stakeholders are now and where they want to be in the future. Especially as the scope of the questions being considered grow beyond the housing and the additional subject matters at play become more specialized and exact, it's important to pivot to strategies of gathering, organizing, and presenting information so that participants and decision makers in the process are speaking from the same set of facts and sharing the same reality.

Geographic Information Systems (GIS) can do just that. It can facilitate a more unbiased platform for information to be studied, issues and trends scrutinized, permutations of various impact assessments to be played out. Alongside the qualitative data, you can then visually articulate and graphically render information in ways that illustrate the findings in the broadest possible terms promoting inclusive engagement and easy digestion of the factors at play.

As a facilitation tool, GIS allows housing discussions to become a collaborative and iterative process where users can draw on 21st- century spatial analysis made more reliable with an ever-growing set of data-rich and accredited inputs. In its simplest form, this is software for locating things on a map, but in the hands of trained professionals, it can set the stage for a proposal to advance more quickly past rudimentary steps and onto the technical and political landscape that needs more attention and nuance.

The possibility that the strategic use of GIS can contribute to affordable housing campaigns and organized movements is more important now than ever. Large cities, and more towns and villages typically untouched by what were considered "urban" problems, now face record numbers of homelessness, increased household overcrowding, and deepening rent burden. The accelerating cost of insurance and materials, rising interest rates, high land costs, and the challenges associated with maintaining quality housing has made conditions for adding new supply more unpredictable for even the most experienced builders. Unsurprisingly, the ramifications go further when you consider that housing shortages can stifle job growth; undermine tax revenue; curb spending on core public services like transit, waste removal, schools, and recreational spaces; exacerbate climate resiliency issues; and dilute fair housing goals and investments to reverse intrenched residential segregation that local, state, and national entities have made. The difficulties not only present issues for diversifying access to housing but it also stiffens obstacles that exist for diversifying the companies working in the sector itself. Emerging and BIPOC builders struggle to break in and overall prevents more firms constructing the housing to reflect some of the communities they are building in.

We are not helpless in this situation. In fact, there are many effective strategies we can deploy to create the housing we need. A suite of tools that include public – private partnerships; social housing strategies; flexible as-of-right and discretionary municipal financial incentives; rental subsidies; permissive and incentivize zoning; and regulatory, code, and administrative reform are among the primary instruments. Federal funding directed at lowering housing costs, expanding supply, improving affordable rental and homeownership options, supporting even deeper levels of affordability, and tackling homelessness creates energy at the highest levels of government that can help break through parochial roadblocks. But to secure these tools requires public support and the key to garnering public support relies on communicating clearly, authentically, and with exactitude – three things that GIS can help us all do.

The authors Ramasubramanian, Albrecht, and Rojas De Leon do a tremendous job working through the complexity of the history, the present, and the future of housing policy decisions at the core of this problem and expertly present the case for how comprehensive spatial analysis can diffuse noise to make room for lucidity in a combative space. My hope is the reader sees this not as a passive learning experience but a call-to-action where every able-bodied and skilled practitioner is compelled into service. The promise of "home" particularly for those with no or limited choice, and the future of our communities depends on it.

**Ahmed Tigani**
*First Deputy Commissioner,*
*NYC Department of Housing Preservation and Development*

# Author Bios

**Dr. Laxmi Ramasubramanian** serves Hunter College as an associate professor of Urban Planning and Policy and the deputy director at the Institute for Sustainable Cities. She holds undergraduate and master's degrees in Architecture from the University of Madras, India, a master's degree in City Planning from the Massachusetts Institute of Technology, and a PhD in Architecture from the University of Wisconsin-Milwaukee. She is also a certified planner holding the AICP credential. She has been recognized as a leader in the geospatial community and has served as the elected president of the University Consortium for Geographic Information Science (2012–2014). In February 2016, she was appointed to a three-year term on the National Geospatial Advisory Committee, a federal advisory committee that provides guidance to the federal government on matters of national geospatial policy.

**Dr. Jochen Albrecht,** GISP, is a professor of Computational and Theoretical Geography at Hunter College and The Graduate Center of the City University of New York (CUNY). This is his third book and the second co-authored with Dr. Ramasubramanian. In addition, Dr. Albrecht published over 60 refereed articles and conference proceedings. He served on the Board of Directors of several professional organizations and is currently the president of the GIS Certification Institute (GISCI).

**Deborah Rojas De Leon,** RA, is a licensed architect in New York. She is the owner and founder of Rojas AP, an architecture and planning firm in Jamaica, Queens. Over the past decade, her work has been dedicated to enhancing the quality of neighborhoods through architecture and urban planning. She consistently seeks out opportunities to collaborate with local community organizations and stakeholders, ensuring that her projects reflect the unique needs and aspirations of the people she serves. Deborah continually works toward a more equitable and harmonious urban landscape for all. She teaches Graphic Communication at Hunter College in the Urban Planning and Policy master program where she empowers future planners to effectively share their spatial stories, enhancing their ability to communicate effectively.

# 1 Why Geography Matters in Housing

## 1.1 HOUSING IS A HUMAN RIGHT

Housing is rooted in the provision of shelter. Along with sustenance to nourish the body, shelter is a basic need. The United Nations Universal Declaration of Human Rights recognized adequate housing as a human right as early as 1948 (UNGA, 1948). According to the United Nations Committee on Economic, Social and Cultural Rights (UNCHR), the right to adequate housing cannot be equated with shelter alone, stating that for housing to be adequate, housing should be suitable for human habitation and include services such as clean drinking water, and sanitation, located away from health hazards or polluted areas, be accessible to everyday activities such as employment opportunities, educational and recreational opportunities, and essential services such as hospitals and health care facilities. Adequacy also includes affordability and security of tenure (Office of the UNCHR, 2009).

Some countries outside the United States as well as some individual communities and cities within the United States are also in the process of establishing policy language that emanates from the housing-is-a-human-right worldview (Fallon, 2021). These policies facilitate an increase in funding to build new or repurpose old infrastructure to produce new shelter options (places to live), in addition to funding support services that are essential to serve those who live in a constant state of housing precarity. While the UN declaration has been accepted by the United States in principle, no federal laws currently exist to protect a right to adequate housing. Recently, a bill called "Housing is a Human Right Act of 2021" (Jayapal et al., 2021) was introduced. While this legislation seems unlikely to receive traction in the US Congress, it calls attention to the need for a serious societal conversation about our national "housing crisis".

## 1.2 THREE HOUSING CRISES

Three distinct housing crises currently plague the United States: the housing supply crisis, the housing affordability crisis, and the homelessness crisis. The housing supply crisis is characterized by an acute shortage of housing units, a longstanding problem in the United States. The affordability crisis focuses on the rising costs of owning or renting a home, making it increasingly challenging for individuals and families to secure affordable housing. Lastly, the homelessness crisis highlights the growing number of people without secure housing, exacerbated by factors such as lack of affordable options and economic downturns. By examining these interconnected issues, we aim to shed light on the complexities of housing challenges. Moreover, it emphasizes the importance of using a geographical lens and utilizing

DOI: 10.1201/9780429243066-1

Geographic Information Systems (GIS) to facilitate conversations and inform policy decisions in the realm of housing.

As we emerge from the global pandemic, the media is drawing attention to a variety of housing challenges. The first framing is that there is an acute shortage of actual housing units for anyone who needs a permanent place to live, in other words, a *housing supply* crisis, see Figure 1.1 (data source, US Census, 2021b). For instance, in July 2022, a *NYTimes* article opened with the claim that "the United States has a deep, decades-old housing shortage" (Dougherty and Casselman, 2022). The claim in this particular article is an attention grabber, at best. How do we know there is a housing supply crisis? Where's the evidence? The rest of this article actually focuses on providing one explanation to address the lack of housing supply, namely, the private market's reluctance to create (build) actual housing units because private home builders remain concerned about their ability to sell homes to credible buyers.

This article and others like it reveal some of the challenges of making housing the subject of academic inquiry. The decision to buy a house is a deeply personal and therefore a very subjective choice and one that is imbued with complex layers of meaning about what it means to become a homeowner. The production of housing, on the other hand, is a business proposition – where a home is commodified, packaged, and sold as an aspirational ideal to millions of Americans who are often enticed to become "first-time home buyers" and hence first-time homeowners, see Figure 1.2 (data source, US Census, 2021b). From a home builder's perspective, the costs of housing production are very high (particularly on the West Coast, the Midwest, and in the Northeast, see Figures 1.1 and 1.4) and there is little room for error, and they typically proceed with caution. The private market does not want a repeat of what happened after the Great Recession of 2007–2009, when many homeowners defaulted on their mortgages and walked away, resulting in a glut of unsold housing. Thousands of units of built housing developments languished for years without occupants (Healy, 2016) as buyers could not qualify for mortgages. Of those who did, many were unable to keep up with the monthly payments, and their homes were repossessed by lenders. In addition, the US economy relies heavily on global and regional supply chains to provide the raw materials and finished products that are critical to housing construction. These supply chains were disrupted because of the global pandemic, another reason that is offered for slower-than-usual housing production in 2021 and 2022 (Goodman, 2022; Sisson, 2022).

The question of whether we have an adequate supply of housing is simultaneously a data-driven inquiry and a philosophical musing because adequacy can be qualified and interpreted in many different ways. In the United States, the dominant form of housing, about 2/3rd of all housing stock, is a detached single-family home, and the mean homeownership rate has hovered between 65% and 66% over the last six decades, see Figure 1.3 (data source, U.S. Census Bureau, 2014 and earlier).

The second framing is focused on the rising costs associated with owning or renting a home, the *housing affordability* crisis. Narratives about the affordability crisis usually focus on homeowners and their challenges of buying a new home. Unlike many parts of the global south, where homebuyers raise most of the purchase price over an extended period, making home ownership very challenging (e.g., Haub and

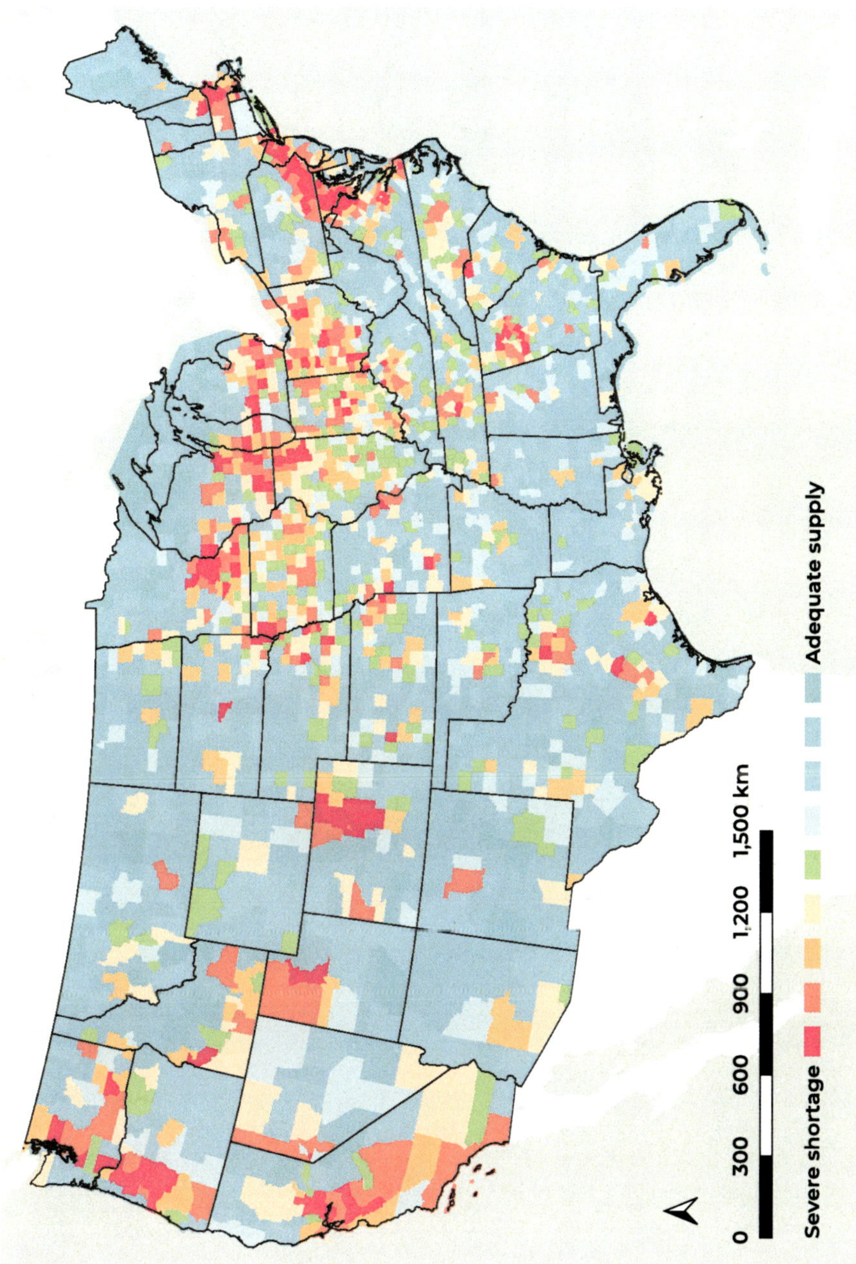


**FIGURE 1.1** Housing supply crisis

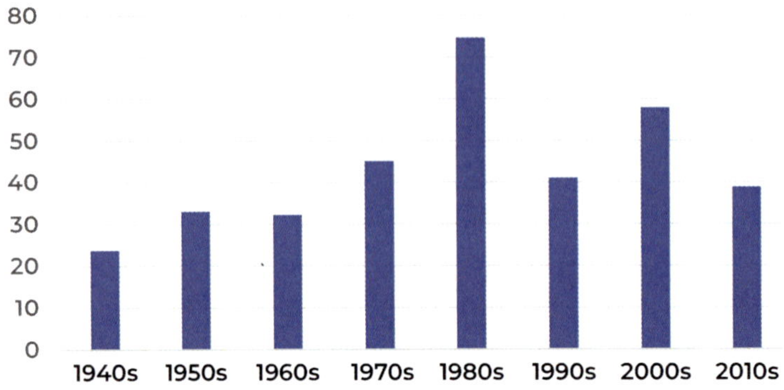


**FIGURE 1.2** Housing built per 100 new residents

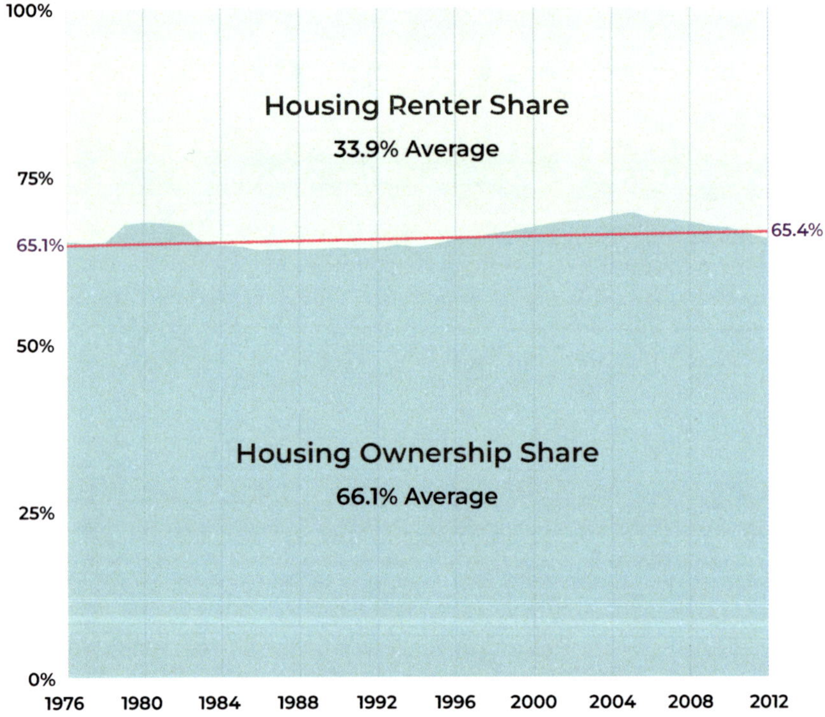


**FIGURE 1.3** US Homeownership trends through the decades

Sharma, 2018), the United States is unique because of the sophisticated financing/lending mechanisms that allow individuals to typically purchase a home with only 20% of the purchase price in hand (Jones et al., 2017). The high reliance on borrowing creates additional instabilities because of the fluctuations in interest rates for home loans. For example, one perspective that is often offered to explain the sales slump in 2022 is that rising interest rates and economic instability make prospective first-time home buyers anxious – they decide to wait a little longer (Kaysen, 2023).

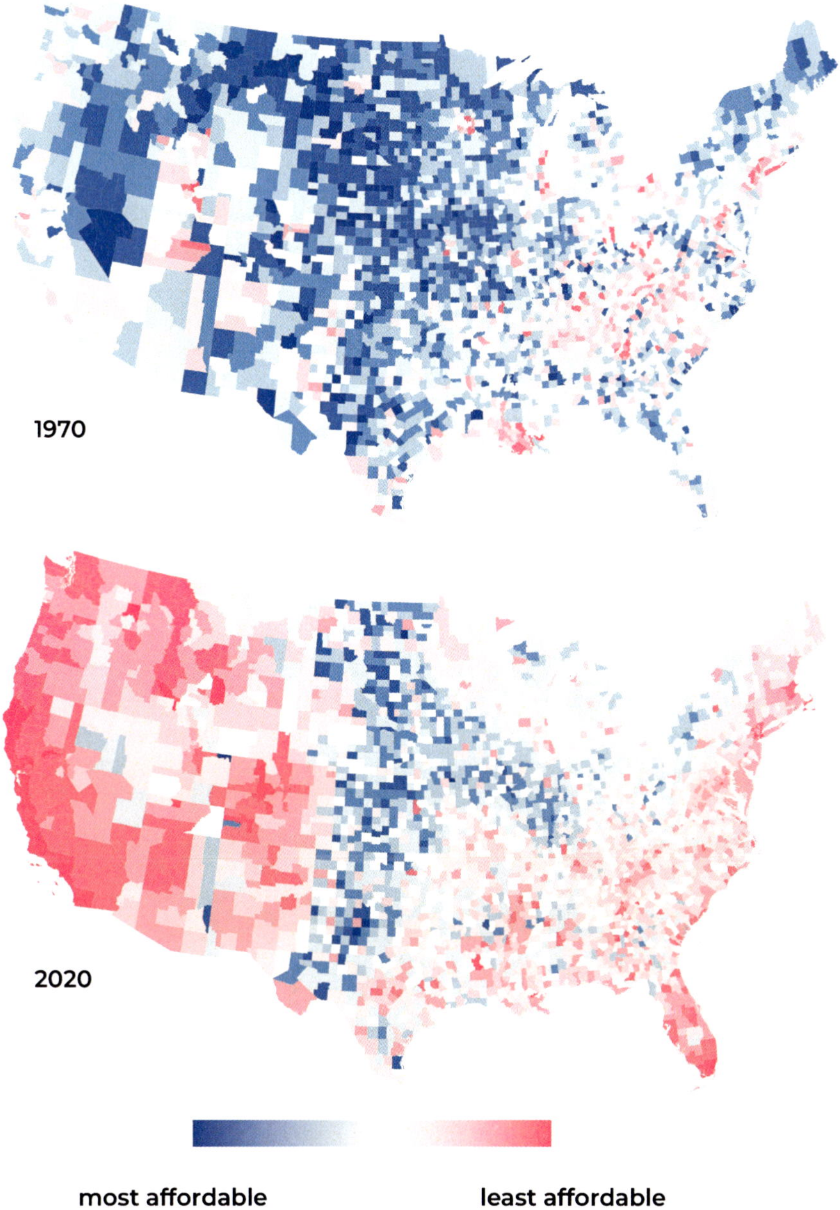


**FIGURE 1.4** Comparison of housing affordability in the United States, 1970–2020

For existing homeowners, the received wisdom suggests that they use the money they have to make improvements on their existing home, rather than buying anew. In urban environments, policymakers and elected officials often discuss the challenges facing renters, another dimension of the affordability crisis. High rents are a source of frustration to many for whom owning a home is impossible, or at best an elusive goal. Renters contend with a myriad of challenges, chief among them is the threat of rising rents that result in displacement. While economists advise individuals and households that they should spend 30% of their income on housing (rent), a typical renter in expensive housing markets like New York City or the Bay Area is more likely to spend between 40% and 45% of their income on finding a decent place to live (e.g., BLS, 2022) – and many renters spend more than 50% of their income on rent (as discussed in some detail in Chapter 6).

The affordability crisis results in *housing insecurity* – a state of instability caused by having to move constantly to find a place to live within a limited budget. Individuals and families on fixed incomes, including disabled individuals unable to work, elderly people relying primarily on social security payments, and workers in service-oriented occupations (schoolteachers, police officers, and firefighters for example), struggle to find stable and affordable housing alternatives whether they own or rent. Housing insecurity is an invisible problem because it can be very hard to assess how close an individual or household is to being evicted. Unstable housing creates new burdens and compounds existing problems being faced by members of the household – children's schooling is negatively impacted, elders may miss out on routine or necessary visits with health care providers, and individual's mental health challenges may be exacerbated because of constant change and anxiety.

The accepted understanding based on the federal government's definition of housing affordability means that a household spends under 30% of the monthly household income towards paying their mortgage (which is the conventional pathway to home-ownership in the United States) or towards their rent (in the case of renters). Figure 1.4 (US Census, n.d. and NHGIS, n.d.) provides the stark visualization of the changes in housing affordability between 1970 and 2020 using counties as the unit of analysis. For reference, there are about 3,200 counties in the contiguous United States, excluding Alaska and Hawaii. The color ramp moves from dark blue to dark red, with the darkest blue color-shaded counties being the most affordable and the darkest red color-shaded counties being the most unaffordable. Counties that are color-shaded white did not experience any discernible changes. Even a cursory glance at the map reveals that housing affordability has decreased throughout the United States in the last five decades. At first glance, housing in several coastal states such as Massachusetts and Florida, in the east, and California, Oregon, and Washington has become more expensive (unaffordable).

While a state may be relatively affordable, regional and sub-regional differences influence an individual or household's experiences of finding affordable housing. In other words, examining housing affordability requires a fine-grained analysis, examining county-by-county variations. In examining changes over time (1970–2020), it may be useful to note that some counties that were relatively affordable in 1970 (dark blue) transitioned to relatively unaffordable. Figure 1.5 (US Census, n.d. and NHGIS, n.d.) shows the details of changes in affordability ranking for counties in California and Nevada between 1970 and 2020. Counties with relatively small populations such as Mariposa county (east of Merced, north of Fresno) show dramatic changes in housing affordability. Mariposa county grew almost 150% in population between 1970 and 2020, but the total population in 2020 was a little under 17,500 people. Housing affordability is a relative measure (the areas that were most affordable in 1970 are expected to show the greatest change in 2020). The value of using counties

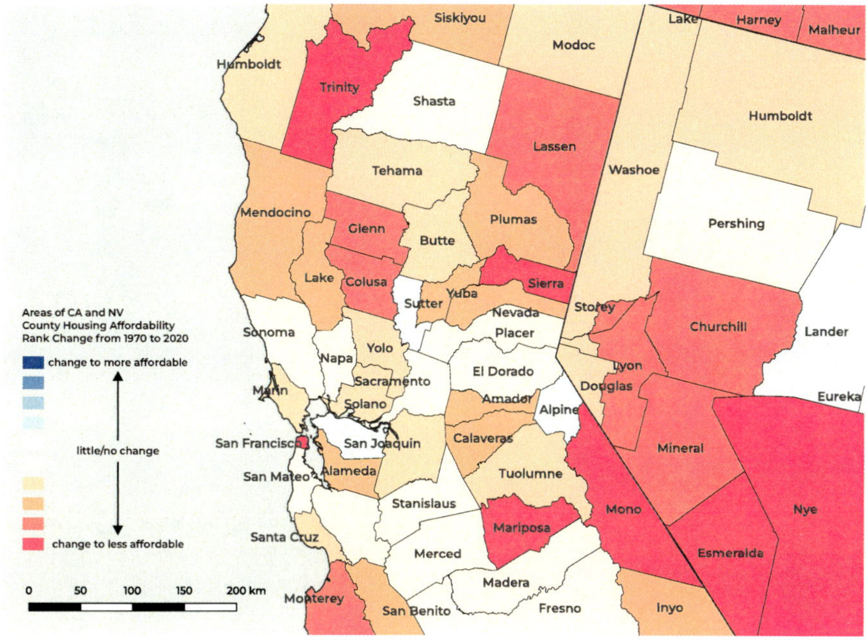


**FIGURE 1.5** Change in housing affordability in California-Nevada 1970–2020

as the unit of analysis becomes more relevant as we see the inter-state dependencies with respect to settlement patterns. During the same time period (1970–2020), Washoe County in Nevada had a net gain in population of over 300%, creating housing unaffordability in that county. Washoe County, includes the city of Reno that has attracted and retained Californians who are able to commute to and from the Bay Area for work and leisure while enjoying the lower living costs in Washoe County, accounting for its rapid population growth. The population explosion places pressures on housing supply, increasing unaffordability.

The economic downturn and the burgeoning public health crises in the United States have resulted in the third and most poignant housing crisis, the *homelessness* crisis. Readers in other parts of the world may be surprised to read that about 0.2% of the US population (a little over 580,000 people) do not have a secure place to spend the night (Meyer et al., 2022). The authors posit that this number is a severe underestimation.

Counting the homeless population is fraught with challenges because of the transient nature of the population being counted. The US Department of Housing and Urban Development (HUD) uses a Point-In-Time count of unsheltered individuals experiencing homelessness in a single night in January every year. This data is combined with other data counts gathered from other non-institutional group quarters such as homeless shelters, domestic violence shelters, and group homes. Figure 1.6 (USHUD, 2022) shows the geography of homelessness in the United States.

One of the main drivers of the current homelessness crisis is related to the lack of affordable housing. According to the National Alliance to End Homelessness (2022), a nonpartisan advocacy organization that tracks and reports data about homelessness,

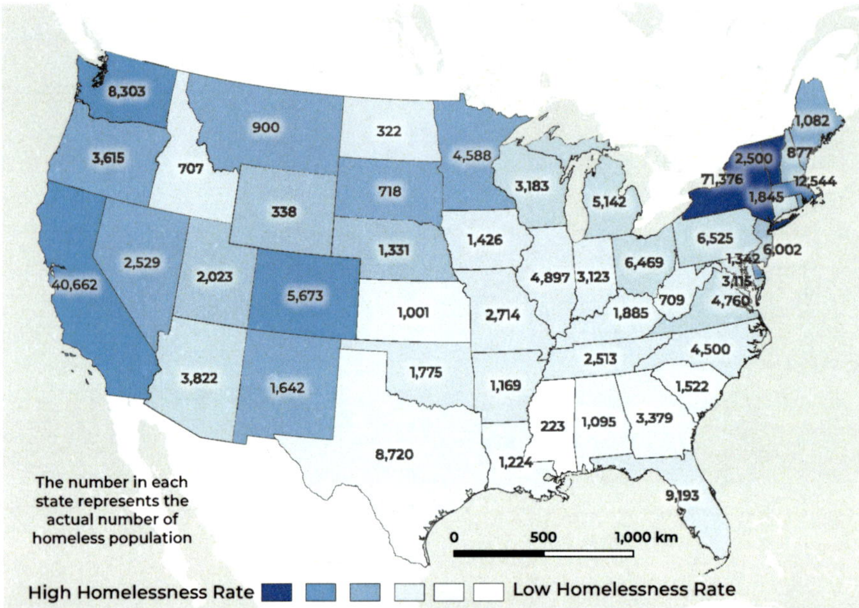


**FIGURE 1.6** Geography of homelessness by state, 2022

unhoused individuals are more likely to be men, about 70% and about 30% of the homeless population comprise families with children. About 8% of the homeless adult population are veterans (Henry et al., 2021).

Unhoused individuals are very often likely to have mental health challenges (National Coalition for the Homeless, 2009). In recent years, young people identifying as LGBT are more likely to be unhoused, although official data is hard to come by since many organizations do not collect this data. Additionally, this information is likely to be volunteered or shared by the individual experiencing homelessness. However, as early as 2011, the National Gay and Lesbian Task Force and the National Center for Transgender Equality identified the challenges of transgender people to find access to safe shelter (Grant et al., 2011).

Once an individual or a household has moved from the ranks of the housed to the unhoused, it is a struggle to return them to their former living situation. This is especially true for those who were already in some form of subsidized housing provided by the state or the nonprofit sector. The visibility of the unhoused provokes a range of emotions in the general public. Although there are frequent demands for action, the problems of the unhoused have become an intractable problem, magnified by waves of asylum seekers fleeing persecution in their countries of origin gathered in southern border cities and towns awaiting formal entry into the United States.

Most scholars who study homelessness discuss the "definitional inconsistencies" as well as limitations of the data (Lee et al., 2010). There is widespread agreement that the numbers of homeless are heavily undercounted because of the invisibility of homeless people and the fact that housing insecure individuals (those who live in their car, couch-surf, or move in and out of the shelter system) often fall between the cracks and may not be accounted for during the single point-in-time count conducted by HUD. For example, the Department of Education statistics is likely to have a more accurate count of school-aged children who live in shelters with a parent or guardian and attend a public school. In recent years, cities in states governed by Democrats have become sanctuaries for people without shelter. Thus, the numbers of homeless people in States such as New York and California with more humane social policies are much higher than states that have criminalized homelessness (Olivet, 2022).

Housing affordability is a social policy question. Authors Donald Schön and Martin Rein (1994) have previously argued that the way policy problems are framed can limit the solution spaces that can be created to address them. They suggested a more pragmatic approach, where attempts to resolve policy controversies are addressed in the context of policy implementation by those individuals or groups that must design and implement the policy decisions through the development of programs. According to Schön and Rein, policy innovations and breakthroughs are more likely to occur as a result of detailed conversations, where understanding different/conflicting policy positions can be fully explored as a part of a pragmatic attempt to solve problems within a specific situational context.

In the United States, the production of housing has largely been left to the market. Housing production, however, is intricately tied to its financing as well as to related infrastructure provision, and all levels of government are involved in creating supportive conditions to allow the housing industry to accomplish the goal of creating housing. Many individual and institutional intermediaries are involved in the supply and management of housing. Housing is a robust area of scholarly inquiry judging by a steady slew of books addressing historical narratives (e.g., Chey, 2017), case studies (e.g., Desmond, 2017), in addition to growth trends and policy critiques (e.g., Madden and Marcuse, 2016) to mention a few. Recently, academic scholars have drawn attention to alternative housing typologies (e.g., Parolek, 2020), as well as unpacking the complexities of creating shelter and rebuilding lives after natural disasters (e.g., Fitzpatrick and Spialek, 2020). Although strong academic linkages have been established between the fields of housing and community development, the academic study of housing continues to be very challenging, because of its central status within the economy, and because of the complex emotional overtones associated with homeownership. There is also a strand of research that links housing challenges within the land use and transportation planning literature (e.g., Jackson, 1985; Kunstler, 1993; Rothstein, 2018).

While data and statistical analysis have been used extensively in the study of housing, the multi-scalar dimensions of these analyses appear to have been curiously neglected by the housing policy studies. To address this disconnect, the authors of this book, representing the fields of architecture, geography, and urban planning strongly recommend the use of GIS tools and the use of a geographical/spatial lens to reframe housing policy debates. Ramasubramanian (2010) and Ramasubramanian and Albrecht (2018) have argued that GIS can make policy conflicts more visible to decision makers and facilitates the rapid testing of different scenarios and policy alternatives that can allow for new policy alternatives to emerge. In a special journal issue that assessed the state of housing scholarship between 1968 and 2008, Galster wrote that, "*academe and the practice of planning and policymaking are like two neighbors, sometimes quarreling, sometimes exchanging resources, always reacting to and stimulating the other*" (Galster, 2008). Agreeing with Galster, we propose that both sides (academics and policymakers) include a geographical lens and use the advanced analytical and visualization capabilities of GIS to facilitate and mediate conversations between housing policy experts, elected officials, land use planners, and community residents to solve housing problems at the neighborhood and sub-regional level.

## 1.3 UNDERSTANDING HOUSING GEOGRAPHIES

Following our claim that housing studies can benefit from using a geographical world view, we draw on the words of Amos Rapoport, architect and author of an influential book *House Form and Culture*, who observed that "*the house is an institution, not just a structure, created for a complex set of purposes. Because building a house is a cultural phenomenon, its form and organization are greatly influenced by the cultural milieu to which it belongs*" (Rapoport, 1969, p. 46). Extending Rapoport's argument, we posit that the study of housing cannot be viewed merely as an assemblage of houses on a street or a neighborhood, housing morphologies and settlement patterns are likely to reflect a dominant cultural ethos that may be as significant or more significant than building with considerations of nature, weather, and climate in mind. Taking a conscious geographical (spatial) view can assist with deciphering those cultural variations and complexities while also helping to delineate unifying ideals.

### 1.3.1 Cultural Beginnings of Early 20th Century City Planning

The original inhabitants of what we now consider the United States of America, the Native American (American Indian) peoples' living environments and settlement patterns were influenced by their own indigenous cultural traditions which were severely harmed by the American settler colonialist project (Hixon, 2013). The earliest European settlers who came to America imposed their (own) cultural norms on the landscape because they considered the place as a tabula rasa upon which they could create their own imprint. The physical settlement patterns came from the cultural landscapes and memories they carried with them and the changes they hoped for as settlers. While the long trajectory of settler colonialism and its tragic impacts on the indigenous communities and landscapes are outside the scope of this book, we respectfully remind our readers that every wave of settlers has shaped and continues to reshape our communities since the beginning (Cavanaugh, 2020).

In our narrative, we fast forward to the City Beautiful movement, a progressive reform movement that began in the late 19th century is a good starting point to understand where the current cultural norms of residential settlement patterns in the United States were established (Hall, 2014). That movement reflects the dominant cultural ethos of the time – establishing the linkages between physical and moral order. The educated and upper classes of the 1890–1900s were concerned about the chaotic conditions in cities – American cities including Boston, New York, Chicago, and even smaller cities like Pittsburgh were overcrowded, dirty,

**FIGURE 1.7** Unsanitary urban areas

and dangerous places. Housing was in short supply and sanitation systems were not well developed. In a bid to reduce congestion and manage overcrowding, the social reformers sought to transform physical places – through legislation and policy.

Figure 1.7 (Rothstein, 1938) is a photograph taken by Arthur Rothstein of a degraded urban environment showing trash, and generally run down building conditions in Pittsburgh in 1938. The actions of the reformers improved the quality of life of urban areas and created many positive outcomes during that time. The ethos of the time emphasized environmental determinism, where disorder in the built environment was correlated with disorder in the social environment. The earliest building regulations were intended to reduce overcrowding and provide for well-ventilated and safe residential living conditions for the urban underclass (Hall, 2014).

The social reformers of the late 19th and early 20th century were not simply content with correcting wrongs and undoing harm caused by poorly constructed and laid out dwellings; they also articulated aspirational ideals for urban living and advocated for governmental and philanthropic experiments to develop new models (Hall, 2014). At the same time, these visions of the ideal city in the early 20th century can be conceptualized as a pragmatic response to the ills of the late 19th century (Ramasubramanian and Albrecht, 2018). One of the dominant visions that were imported to the United States included the Garden City concept advocated by Ebeneezer Howard in 1898 and 1902 (Howard, 1898/2010; 1902/1965). The concept was often described as accomplishing a balance between city and country living. The balance was partially accomplished by separating functions that did not belong with each other – places for living (residential areas) were consciously separated from places associated with work (industrial areas). Because the work at the time was often noisy, dusty, and sometimes dangerous, these arrangements seemed plausible. Figure 1.8 (Howard, 1898) shows the elements of the Garden City concept that relied on rail transportation to connect human settlements of modest scale/density surrounded by agricultural uses. The segregation of the sick and the elderly (convalescent homes), and people with physical disabilities (asylums for the blind and deaf) visualized indicate elements of the concept.

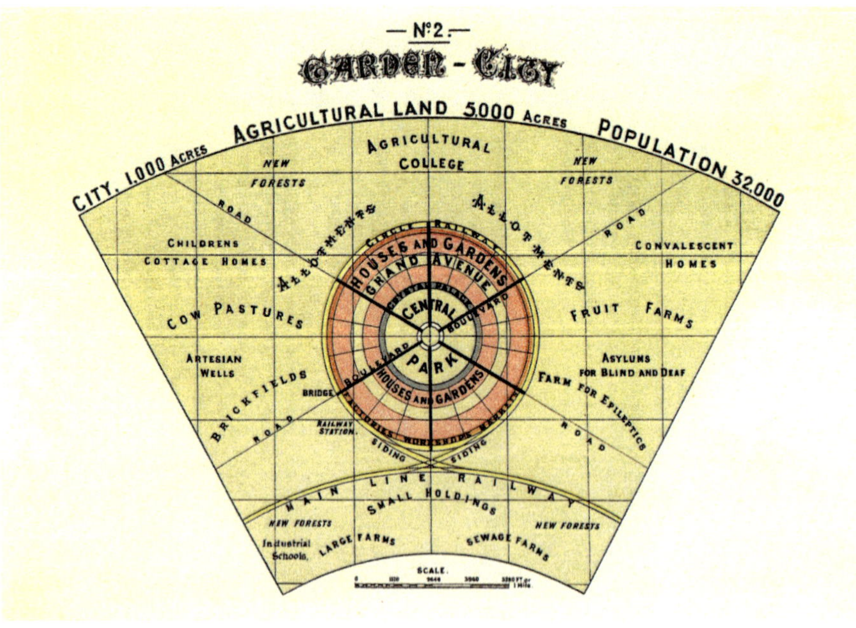


FIGURE 1.8 The garden city concept

### 1.3.2 Dominant Models of American Urbanization 1900–1945

Academic scholars aligned to explain the confusing and complex morass of American cities by beginning the process of codifying the internal structure of the city. Scholars such as Park and Burgess from the Chicago School began modeling city morphologies in abstract terms in the early 1920s. The dominant ethos of the time viewed the American city as a biological milieu (a social-ecological view) that used the language and reasoning of how natural environments thrived and evolved to social and community environments like neighborhoods. The sociologists of the Chicago School imposed their own cultural biases in describing, explaining, and later predicting how cities were growing and how they would grow in the future. In their conceptualization, see Figure 1.9 (adapted from LeGates and Stout, 2019), the business functions were better suited to be in the center (privileging commerce), surrounded by a transitional zone, followed by three rings of residential housing, moving from dense to less-dense development. Of note is the zone of better residences, which appears to be at a "reasonable distance" away from the business center but yet not so far away as

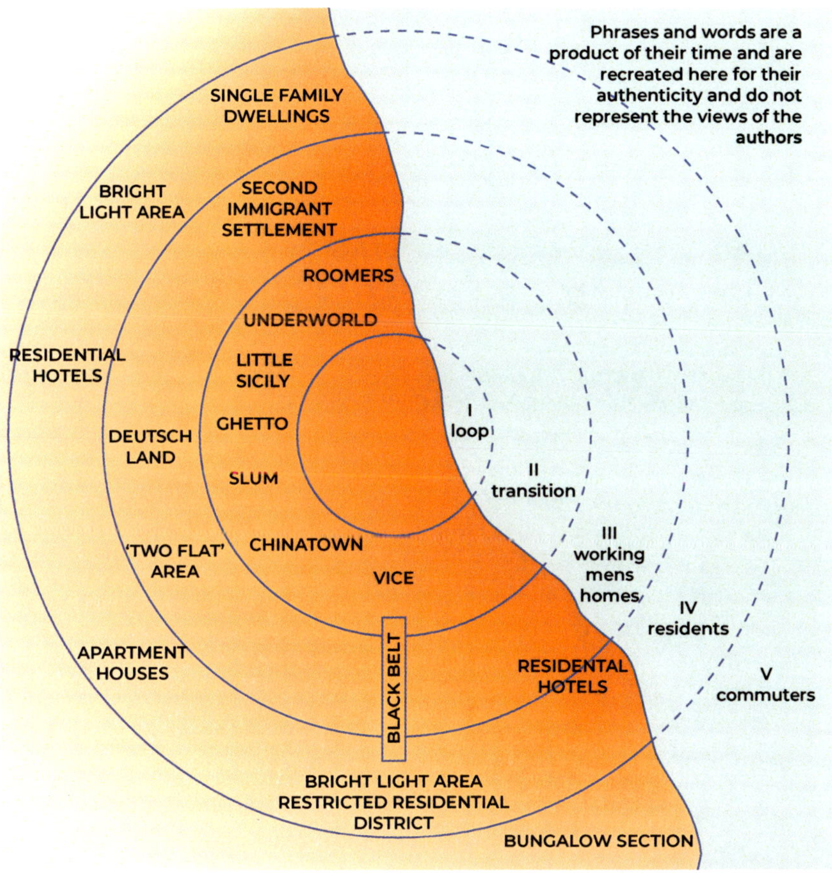


**FIGURE 1.9** The structure of the city conceptual model by Park and Burgess. Adapted from LeGates and Stout (2019)

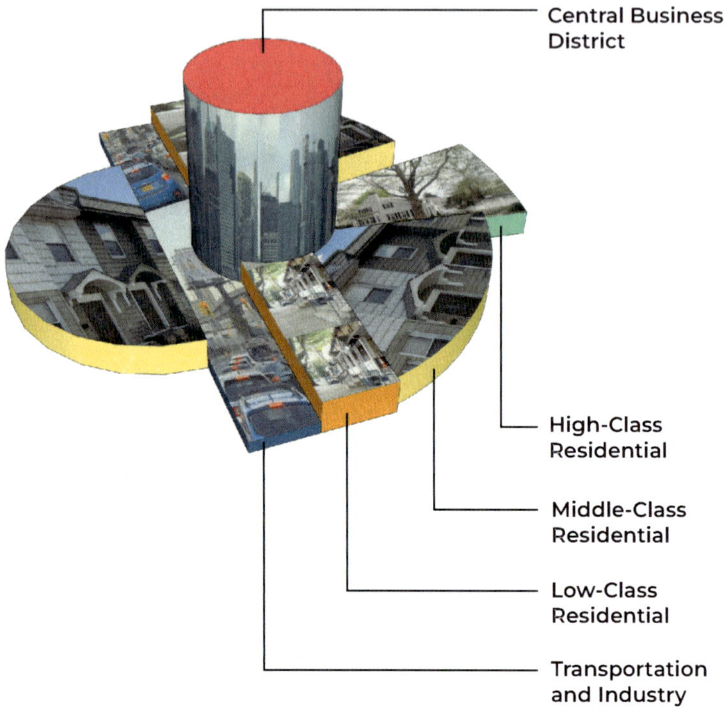


**FIGURE 1.10** The structure of the city conceptual model by Hoyt. Adapted from Hoyt (1939)

to be a non-city dweller or commuter (Park et al., 1925/1967). The "concentric zone model" can be adapted to accommodate the realities of natural topographies and the constructed realities of accessibility corridors such as railroad routes.

Subsequently, additional models to explain city living were proposed, including by Hoyt (1939) who proposed that the cities evolved as sectors; here the high-end residential sector moved along a predetermined transportation corridor/routes such as a streetcar or train line whereas the lower-rent districts were more likely to be adjacent to industrial areas or freight corridors. In both models, preferred residential areas are likely to be segregated from "noxious" uses/activities, whatever those activities may be. Distancing and spatial segregation of uses as a way of commodifying and adding value to certain residential areas were established early on. Therefore, any discussion of housing geographies must be linked to a consideration of neighborhood geographies. See Figure 1.10 (adapted from Hoyt, 1939).

### 1.3.3 Neighborhoods and Urban Settlement Patterns

The reformist goals to reduce density and create safe and well-ventilated living spaces, combined with the availability of land, alongside the evolution of new transportation technologies (street cars, passenger trains, and private automobiles), encouraged the move away from inner city living towards proto suburban environments away from the city center and the creation of predominantly residential neighborhoods within cities. The "neighborhood" as a spatial and social unit is a persistent idea that has occupied the geographical and social imagination for over a hundred years and remains curiously unchanged since its original conceptualizations. The identity of neighborhoods in these cities is deep and well developed since the 1920s. For example, Perry (1929) attempted to define a neighborhood unit, in part because architects and planners laying out new residential areas for the growing population needed a way to organize them – to provide services, to market the new areas to prospective buyers, and to facilitate and manage orderly development and growth. Perry specified areas with distinct boundaries so that residents could visualize it as a distinct entity to accommodate between 3,000 and 6,000 people (ref). Shared services included schools, playgrounds, and parks to be located within the neighborhood, while shops and commercial areas were located along the peripheries, see Figure 1.11 (adapted from Perry, "The Neighborhood Unit", 1929).

According to Glass, 1948, cf. Walmsley and Lewis (1993), a geographical neighborhood provides a means of translating social distance into geographical distance, affords a convenient unit for the provision of goods and services, and facilitates the formation of a territorial group, in which the members can meet on common ground for both spontaneous and organized social activities. Although the idea of neighborhood unit has been criticized since the 1950s for its superficial understandings of homogeneity and community formation (proximity does not always induce conviviality) and because of the tendency of developers and speculators to artificially shape homogeneity using restrictive and racist covenants, the neighborhood unit became the basis for planning and remains so, a hundred years later. In contemporary terms, a neighborhood is an imagined place, one that is a relatively homogeneous and cohesive agglomeration of residential living units that share some basic services and amenities but one that is simultaneously distinct and disconnected from other such agglomerations.

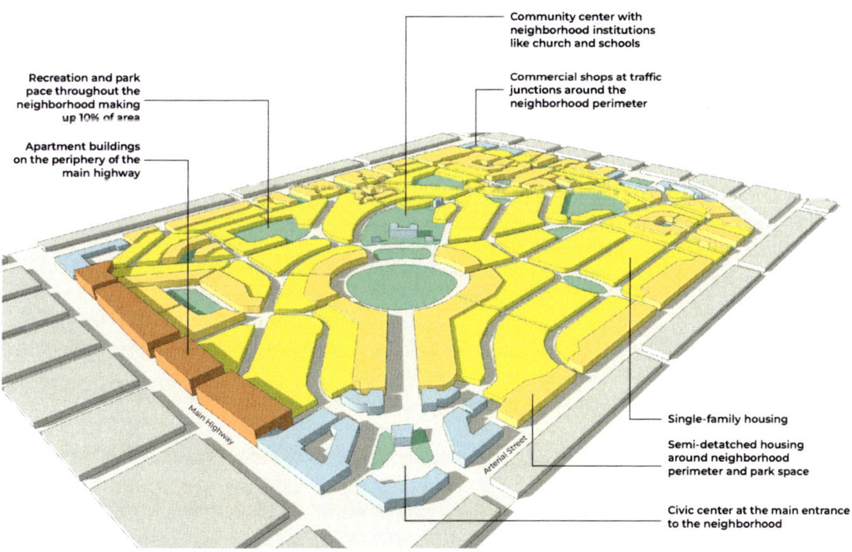


**FIGURE 1.11** Clarence Perry's conceptualization of a neighborhood unit. Adapted from Perry, "The Neighborhood Unit" (1929)

As city planners began codifying activities/uses that could occur within neighborhood units, determining what amenities should be included in each neighborhood (for example, parks and playgrounds), and what amenities should be shared among neighborhoods (for example, shopping), they also began to formalize the separation of activities and uses. New York and San Francisco were among the first cities to establish zoning ordinances that created "districts" or "zones", designating large areas as "residential". Fisher (1962) reviewing the San Francisco experiments writes:

> Whereas a building code emphasizes considerations of structural and fire safety, and a housing code focuses upon those features of a dwelling unit that make it decently habitable, a zoning ordinance is more concerned with the integrity of a neighborhood as a desirable place in which to live or work. As such, it is an essential element of a city's program for the preservation of existing neighborhood values and the guidance of future development. The essence of a zoning ordinance is its designation of separate use districts for the three broad categories of residential, commercial and industrial uses.
>
> *(Fisher, 1962: 326)*

The gradual shift from legislating the form and function of individual buildings to legislating the form and functions of a neighborhood had a significant impact on Americans' cultural understandings of housing.

### 1.3.4 Suburbanization and Suburbs

Advances in transportation technologies, primarily the private automobile, are often credited with encouraging and stimulating the first waves of suburban development. In New York, for instance, Robert Moses established the scenic parkways that would lead affluent New Yorkers to northern and western suburbs as early as 1925 (Caro, 1974) away from a crowded and noisy New York City into the bucolic garden suburbs. The largest impetus of suburbanization occurred after 1945 when a confluence of public policy decisions supported the movement of newly returning War veterans with opportunities to create the American dream (e.g., Beuregard, 2006). Levittown, Long Island, often referred to as America's first suburb was a vast tract of mass-produced houses, only possible because of the deployment of Fordist models of assembly-line production, and the alliance between the private developers (Levitt and sons) and the federal government (housing loans and guarantees for purchasers). The design of the Levittown homes emphasized a specific kind of residential living centered on the nuclear family that lived in a private enclave, surrounded by the new wealth and luxury afforded by technology (Kelly, 1993). The Levittown single-family home used emerging labor-saving devices and a garage, the designated space to store the private automobile (Gans, 1967, 2017), see Figure 1.12 (Levittown, n.d.) to appreciate the new morphology.

The design of suburbs also gave rise to unique neighborhood morphologies. Suburban neighborhoods were created by subdividing a large parcel of land into a series of smaller lots or parcels to create residential neighborhoods. The older suburbs were created on land that was previously used for another purpose, and the land use designation was changed from its original use (for example, agriculture) to its new use (usually residential). Subdivision processes are regulated by state and local laws and subject to environmental review in some states. Because these are purpose-built allotments of land upon which new houses are built, all at the same time, considerable pre-planning occurs to

**FIGURE 1.12** Levittown

create a pleasing and efficient road layout that links the individual lots, providing safe ingress and egress from the subdivision to the larger highways that will ultimately connect the subdivision to other parts of the urban network. The street layouts in subdivisions are intended for automobile travel and have unique interior street layouts – loops, cul de sacs and curvilinear driving paths, all intended to create a sense of enclosure and belonging for those who are fortunate enough to live there, see Figure 1.13.

The design of the earliest suburbs harkens to Ebeneezer Howard's visions of a "garden city". Southworth and Ben Joseph (2003) discuss one of the unique street features of a suburban landscape – the cul de sac (a dead-end street) design. They argue that the cul de sac street has some benefits to residents because it creates safe streets for children and pedestrians, promotes social interactions among neighbors, and reduces infrastructure development costs for developers/subdivision planners.

Reviewing the literature, Ann Forsyth (2012) codified the dimensions across which a suburb may be recognizably identified, as distinct from a city, including its location within a metropolitan area, its built environment characteristics such as having low-density detached houses, transportation access (car reliant), activities (single uses such as residential only), and sociocultural characteristics of the residents and neighborhood.

Sprawl, the rapid growth of low-density, low-rise residential development in land that was formerly not designated for residential use is a direct result of federal policies that promoted the growth of suburbs after World War II (Hayden and Wark, 2005). The spatial segregation between residential and non-residential uses was only possible because of the

**Traditional Developments: Des Moines, Iowa**

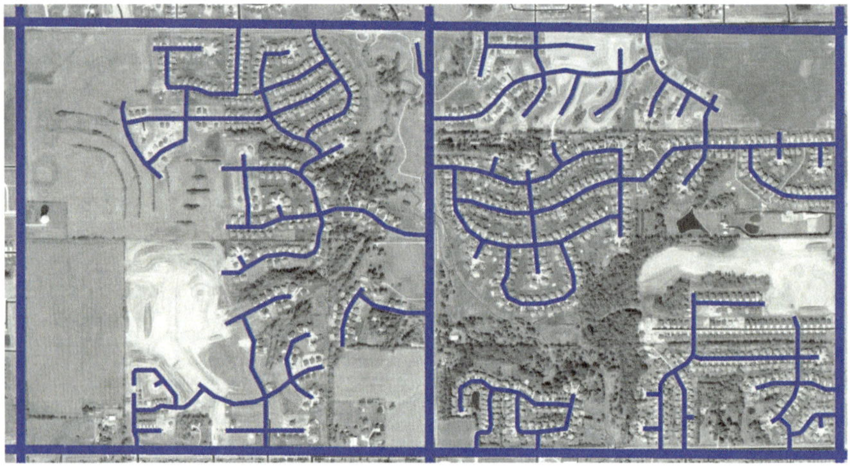

**Cul-De-Sac s Developments: Des Moines, Iowa**

FIGURE 1.13 Traditional grid vs. cul-de-sacs in Suburbia

growth of private transportation and highway infrastructure that increased mobility and accessibility. It is also important to note that both the private sector and the government perpetuated racist policies and actions that have shaped the housing sector – Black veterans could not gain access to the favorable loans offered to Whites, and housing developers explicitly included restrictive covenants that barred Blacks and other minorities from owning a home in the newly emerging suburban developments (Rothstein, 2018).

Although the characteristics of the earliest suburbs are very different from the modern car-oriented suburbs that were built almost a hundred years later, suburbs established the single-family home as the dominant housing form in the United States (Jackson, 1985). While a critique of suburbanization and suburban housing are outside the scope of this book, we want to remind readers that the American suburb is a cultural idea, an imagined

**FIGURE 1.14** Suburban dystopia

place immortalized as a beloved place to grow up on television series such as *The Wonder Years* (1998–1993), while simultaneously vilified as a dystopia in movies like the *Truman Show* (1998) and *American Beauty* (1999), see also Figure 1.14 (Flickr, 2009). It can be argued that suburbs emerged with the support and endorsement of federal, state, and local governments and the enthusiastic support of the real estate industry (Burchell et al., 2005).

### 1.3.5 Small Towns and Rural Settlements

In a discussion of housing in a highly urbanizing world, it is often easy to ignore small towns and rural settlements. There is a great diversity of small towns with a population between 5,000 and 25,000 people in the United States. Some of these towns thrive because of their location – in commuting proximity to job centers or because they have an anchor employer (e.g., a university) that supports the town. However, there are other towns and rural communities that have pressing housing needs because they do not have locational advantages, or they have lost their major employer. While stable housing can be an anchor to good education, healthcare, and employment, conversely, areas with substandard housing are associated with limited or no access to schools, hospitals, or work. Planning for the housing needs of vulnerable populations in sparsely populated small towns and rural areas requires different approaches than what is commonly done in dense urban areas. In the map of Texas, represented in Figure 1.15, there are 104 counties that have no urbanized area at all. The total number of residents in those

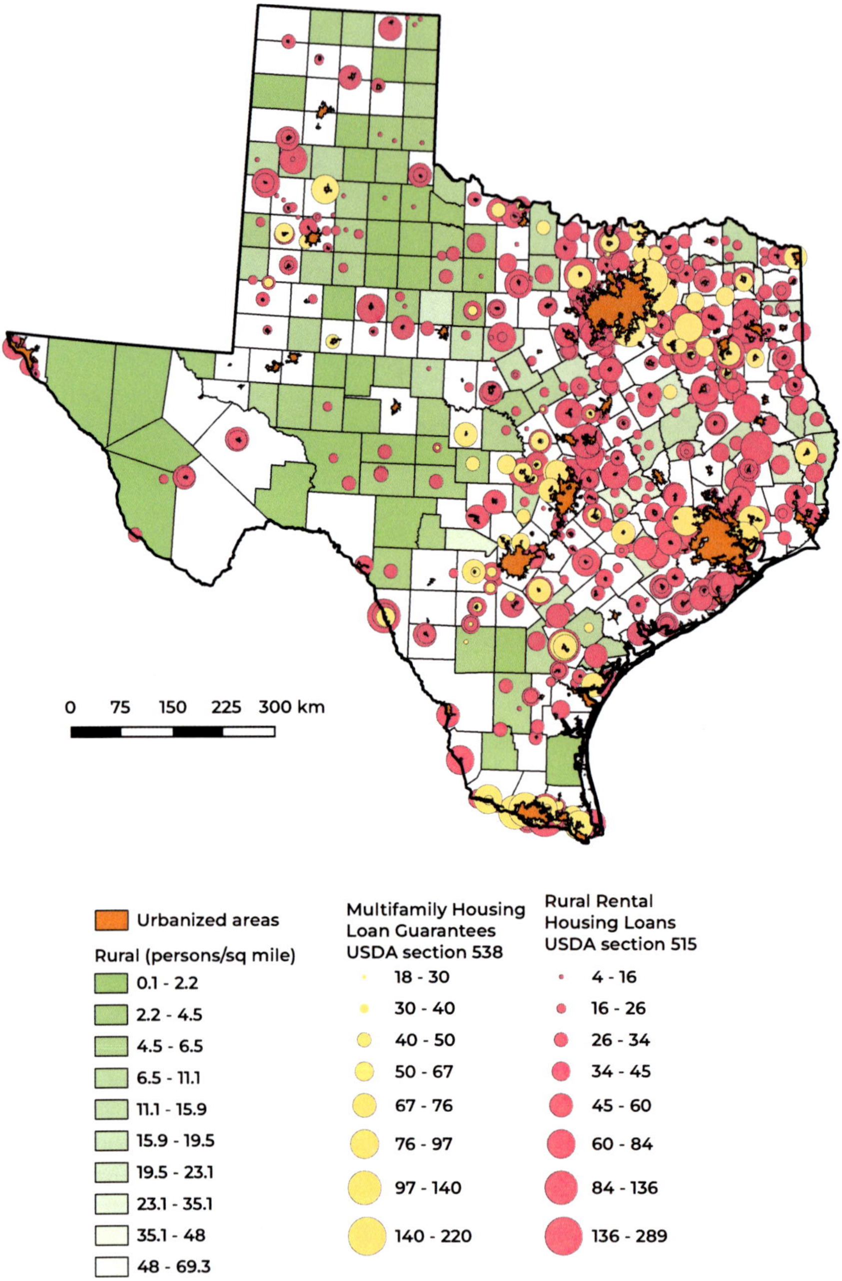


FIGURE 1.15 Rural vs. multi-family housing loans in Texas

counties is some 769,000; as a matter of fact, 66 of those counties have fewer than 10,000 residents. Compare this to Harris county (which includes the city of Houston), where we find 4.7 million residents in a single county. The rural/urban divide in Texas is massive: six counties each have more than a million people and the same six counties also have a population density of more than 1,000 per square mile.

The lack of density in rural counties is the main cause for the lack of infrastructure because it becomes expensive to extend sewer, water, and electrical lines to serve a relatively small number of people. Distances are larger and the tax basis is lower, which results in many counties not making allocations to provide subsidized rental housing, such as those provided by the USDA's section 538 and section 515 rural rental housing programs. As Figure 1.15 (USDA, 2022) shows, rural rental housing tends to be closer to the urbanized areas, leaving many rural counties without adequate rental housing and the poorest rural families such as migrant farm workers without access to adequate shelter, see Figures 1.16–1.18 (US Census, 2021) as a snapshot comparison between a Swisher county, a rural county on the western border and Dallas county.

One of the challenges of rural depopulation is that the vulnerable populations left behind are likely older, sicker, poorer, and less educated than those who left the area. The shortage of rental housing further exacerbates housing affordability and housing insecurity in rural areas (NRHC, 2014).

## 1.4 DEMOGRAPHIC TRENDS AND IMPLICATIONS FOR HOUSING

There is a close relationship between demographic trends and housing. It is well established that the population of the United States is growing steadily, even though the pace of growth has slowed since 2000. The 2020 population was listed as 331 million in 2020, a 7% increase since 2010. Most US states gained population with the exception of West Virginia which recorded a 3% decrease between 1950 and 2020. States like Utah and Texas gained the most people. New births are partially responsible for population

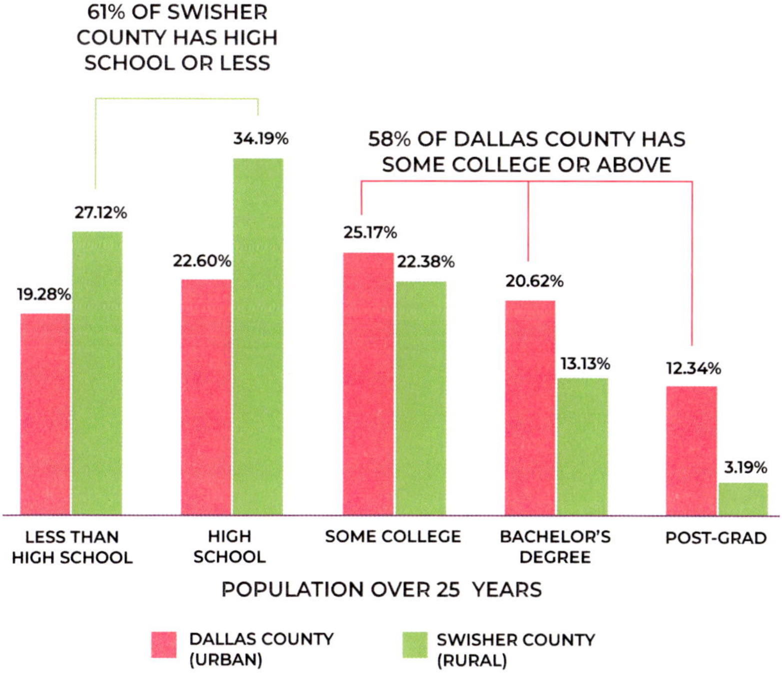


**FIGURE 1.16** Comparison of education level, rural vs. urban in two Texas counties

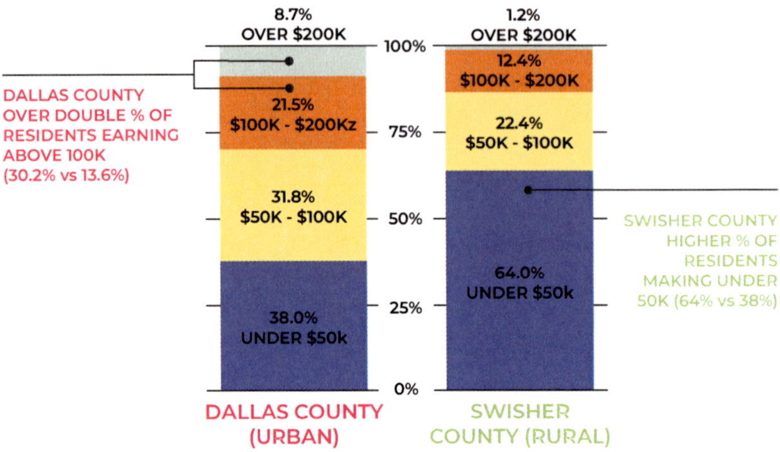


**FIGURE 1.17** Comparison of income level, rural vs. urban in two Texas counties

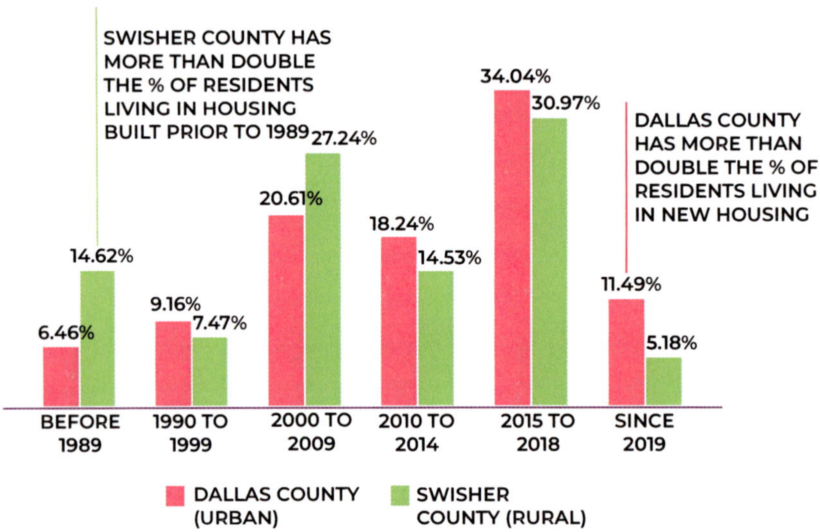


**FIGURE 1.18** Comparison of age, rural vs. urban in two Texas counties

growth, but immigration is another factor contributing to the growth of the US population, projected to closer to 370 million by 2051 (Congressional Budget Office, 2022). See Figure 1.19 (US Census, n.d.; NHGIS, n.d.) for a geographic representation of the percentage of population change in the United States from 1950 to 2020.

### 1.4.1 Housing an Aging Society

While immigrants to the United States tend to be younger and more likely to have children (Frey, 2019), the population in America is aging, see Figure 1.20 (data source, US Census, 2021). Starting in 2030, older Americans (*aged 65 and older*)

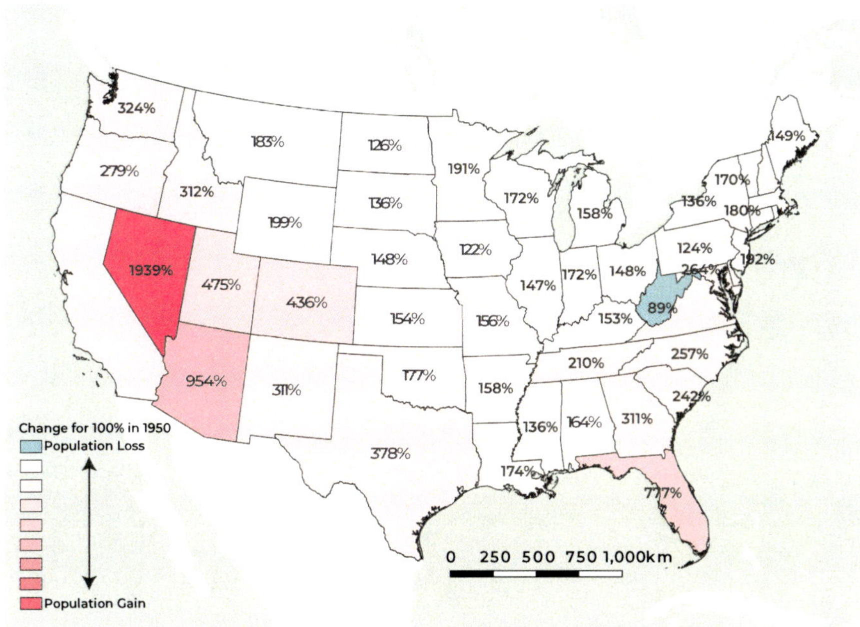


**FIGURE 1.19** US population change between 1960 and 2020 by state

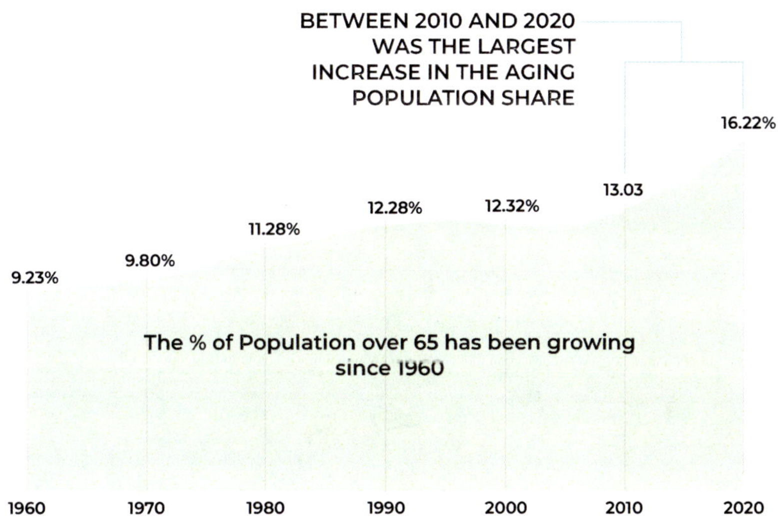


**FIGURE 1.20** US aging population, 1960–2020

will make up over 20% of the total population, increasing to about 25% by 2060. Presently, more than half of all seniors above the age of 65 live in nine states led by California (5.8 million) and Florida (4.8 million), see Figure 1.21 (US Census, 2022).

The American family structure has long segregated its elders – older adults seldom live as part of a multi-generational household. Most Americans express a desire to age-in-place, holding onto their established patterns of living, surrounded by their social

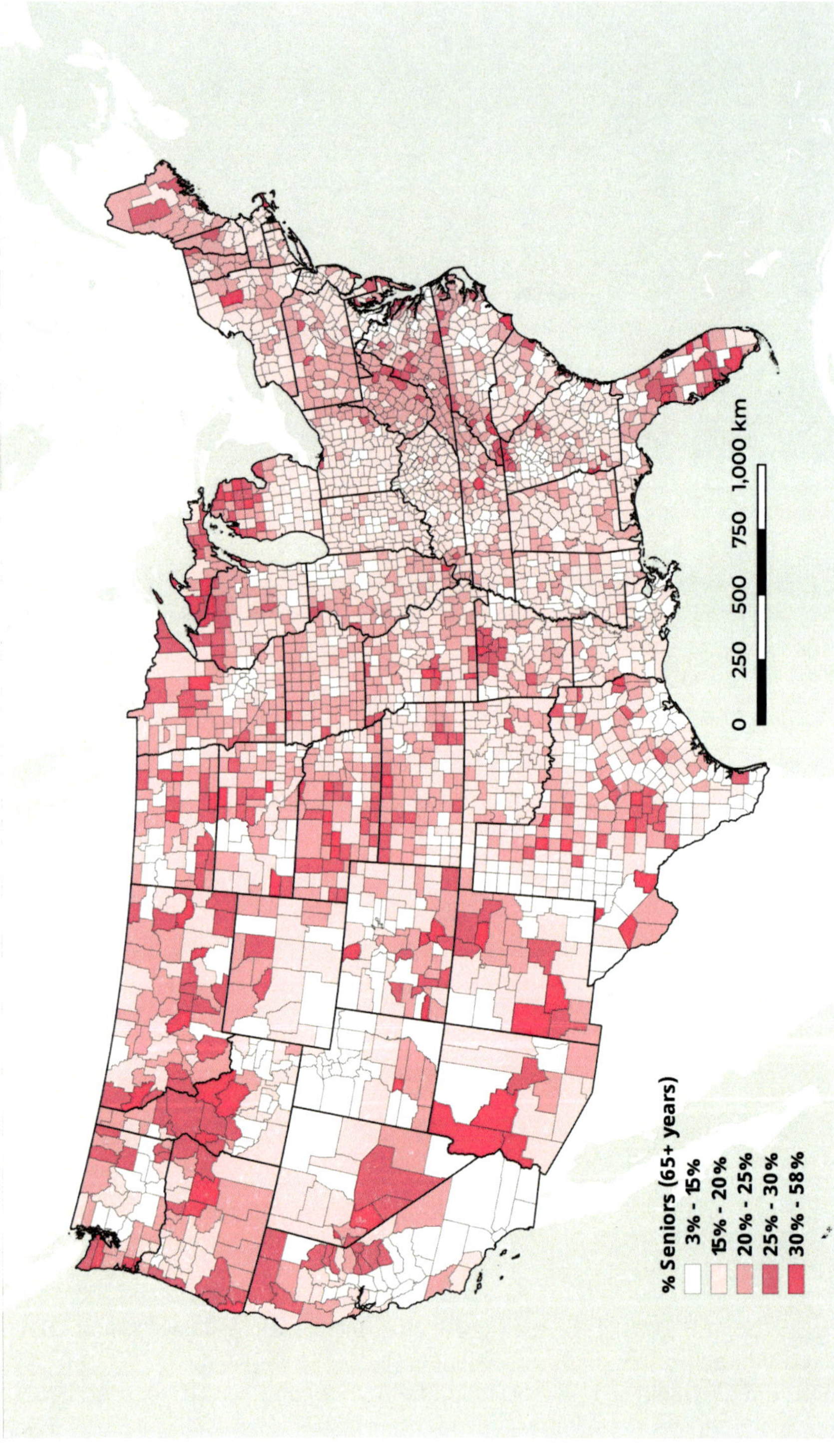


**FIGURE 1.21** US counties showing the presence of seniors

and cultural networks, an aspirational ideal accepted in policy and practice (Means, 2007; AARP Research, 2018). However, as they age, older adults seek alternative living arrangements that accommodate their financial means and their physical capabilities, likely opting to move away from their suburban home. The first step along a continuum of care is usually a retirement (55+) community, also referred to as an age-restricted community. These retirement communities can include a range of housing options, including modified single-family homes, town houses, or apartments. Most retirement communities include a range of support services necessary for successful aging. Age restricted communities can include detached single-family homes, semi-detached residential units with 2–4 units per building, or apartments/condominiums that include 5+ units per building (Foundation for Community Association Research, 2017).

Environmental gerontologists have long argued that the “fit” between an older individual’s individual capabilities and their living environment is central to their sense of wellbeing and a contributor to successful aging (Lawton and Nahemow, 1973). Although elderly residents may not be fully employed in paid work, successful aging models can include both paid and volunteer work, and opportunities to engage in social and recreational activities alone or in the company of others. The presence of “third places” (Oldenburg, 1989) that support opportunities for social engagement appears critical to the wellbeing of older adults (Sugiyama et al., 2022).

Creating and maintaining affordable senior housing is expensive – it is often supported by HUD Section 202 grants or the Low-Income Housing Tax Credit (LIHTC) allocations (Congressional Research Service, 2023). For older adults who have experienced traumatic situations earlier in life, such as homelessness or housing discrimination, researchers (Canham et al., 2022) argue that protections against displacement may be critical to their sense of wellbeing. There is some preliminary evidence suggesting a correlation between housing unit type and perceived social isolation among senior housing community residents, based on a sample of 1,160 individuals living in HUD Section 202 communities in metro Detroit, Michigan. The research found that “*individuals living in townhome-style dedicated senior housing were at lower risk of experiencing social isolation than their counterparts who lived in apartment style buildings*” (Carbone et al., 2022, p. 897).

When we combine geographic (Section 1.3) and demographic (this section) trends, we find that there is a huge difference between elderly populations in rural vs. urban areas. Some inner cities have reversed the late 20th century trend of depopulation and are attracting empty nesters, who cherish the high levels of accessibility, be it to health care or cultural amenities. This contrasts with the classic hinterlands of upstate NY, Appalachia where senior citizens are bereft of services (see also our discussion of NORCs in Chapter 6). Senior residents play an outsized role in rural areas. Table 1.1 is based on a comparison of the 104 Texas rural counties that we presented in Section 1.3.5 with the top six urbanized counties (Dallas, Harris, Tarrant, Bexar, Travis, and Collin).

### 1.4.2 The Challenges of Housing Immigrants

While the desire to move to the United States and become a citizen has remained generally consistent (the United States is still seen as a desirable place to live permanently),

**TABLE 1.1**
**Characteristics of Senior Populations in Rural vs. Urban Counties in Texas**

| Variable | Rural | Urban |
|---|---|---|
| Share of area population | 18.6% | 22.2% |
| Disability | 41.5% | 32.3% |
| Own their place | 86.1% | 75.6% |
| Rent their place | 13.9% | 24.4 |
| Still pay mortgage | 14.2% | 35.7% |
| Without phone | 1.8% | 1.3% |
| In poverty | 9.2% | 7.5% |
| Local for >30 years | 36.6% | 26.5% |
| Veteran | 15.1% | 15.6% |
| Share of area wealth | 36.2% | 29.0% |

the number of immigrant visas issued (legal pathways to permanent residency and eventual citizenship) fluctuates as part of American foreign policy. At the time of writing (2023), immigration from Mexico tops the list with over 40,000 immigrant visas issued in 2021 (Koop, 2022). Legal immigrants to the United States are more likely to arrive at gateway cities such as New York and Los Angeles. Many of them have limited resources and struggle to navigate life in expensive housing markets. Immigrants arrive in the United States for a variety of reasons, some in search of economic opportunity while others flee oppression and persecution. However, it is a truism that all immigrants seek work. Legal immigrants often seek skilled work and regardless of where they arrive, they move to places where they can find meaningful employment. Looking at data between 1970 and 2020, there are no discernable settlement patterns; legal immigrants are found in all 50 states, although the majority are found in Texas, California, and Florida, see Figure 1.22 (US Census, n.d. and NHGIS, n.d.).

We can speculate that immigrants are more likely to become homeowners because they view the pursuit of homeownership as a legitimate pathway to wealth creation. Myers and Pitkin (2013) argue that the share of new homeowners who are foreign born is nearly eight times greater than what it was during the 1970s. It seems reasonable to conclude that as American society ages, adults 65 and over are likely to sell their suburban single-family homes to new immigrants who are younger and are active in the workforce. However, not all immigrants will become homeowners, and even those that aspire to become homeowners will be renters when they arrive in the United States. Salz (2007) investigated the question of how immigrants affect the housing market and observed that there is a local economic impact of immigration, pushing up demand for housing in destination areas, resulting in increased rents. His findings reinforce earlier research by Muller and Espenhade (1985) who observed that rents went up in Los Angeles, more than other metro areas in the period 1967–1983, a phenomenon attributed to the arrival of new immigrants. A similar correlation was observed in Toronto, Canada, by Ley and Tuchener (1999).

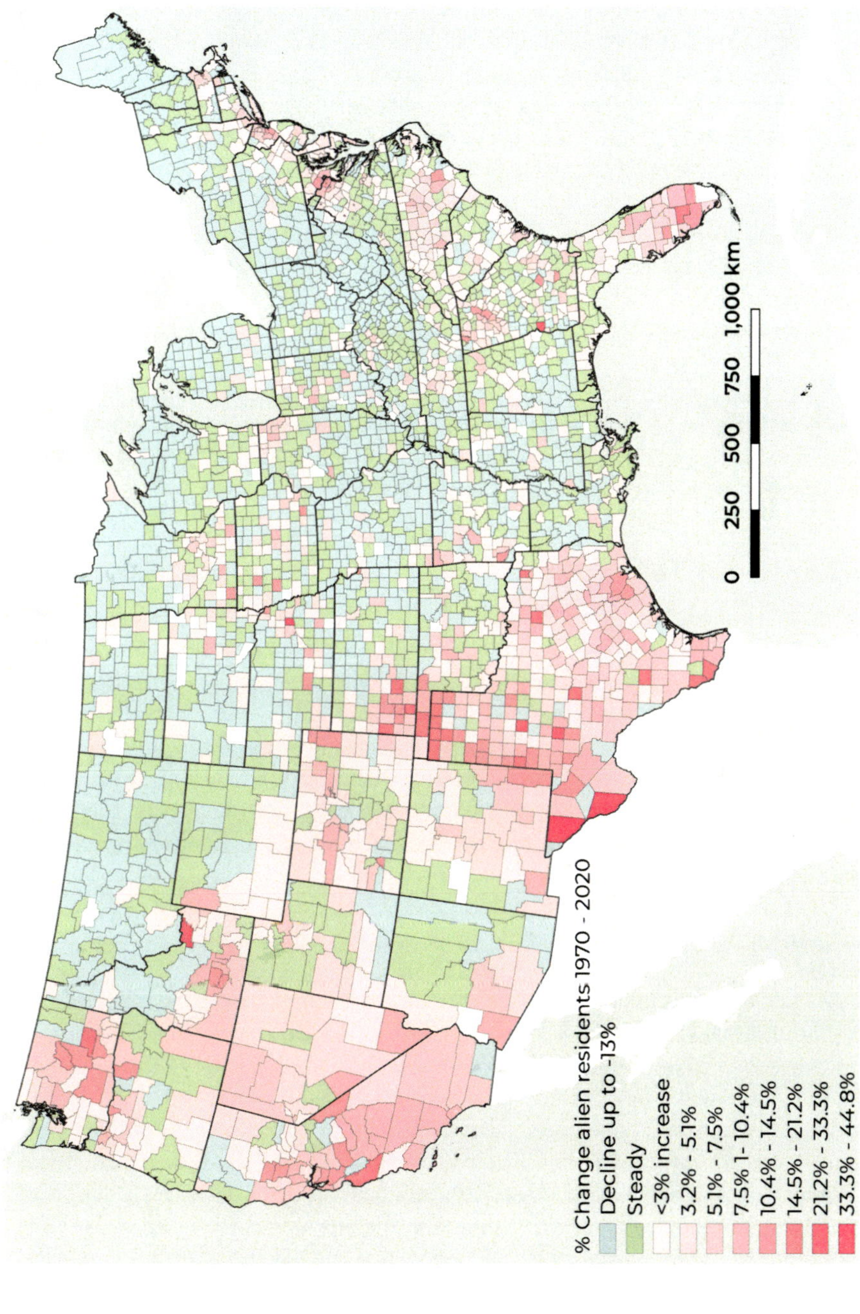


**FIGURE 1.22** Change in immigratior., 197C and 2020

There are other cultural variables to be considered. Many larger US cities have ethnic enclaves, with quaint references to Chinatown, little India or little Saigon. These enclaves are often the beating heart of the immigrant community. While outsiders may consider these enclaves as restaurant districts, for poor immigrants, especially those without formal education, these neighborhoods are live-work spaces, providing much needed shelter in addition to a robust social support system until they can establish themselves. Thus, phenomena such as illegal subletting of apartment units and overcrowding may attract the attention of both the planning authorities and law enforcement. We must recognize that the new immigrants, predominantly people of color, are challenged in the same ways as the predominantly European immigrants who experienced discrimination in the pursuit of jobs and housing in the 1900s. The challenges of securing housing forces immigrants to consume less housing, although we can speculate that over a period of time, their housing needs (in terms of space usage) become comparable to the locals.

Migrants who have arrived in gateway cities illegally are being subjected to extreme hardships by law enforcement. While conditions for asylum seekers and refugees are marginally better, these individuals and their families including young children are pushed into overcrowded shelter systems without much support. Migrants' access to safe and adequate housing is proving to be a major challenge.

### 1.4.3 A Tale of Two Cities

Our visualizations of change over time tell a story of how changes in housing are related to accompanying urbanization and suburbanization trends, demographic shifts, technological advances, economic fluctuations, and politics all affecting housing geographies as discussed in Section 1.3. While it may be easy to label cities and regions as "winners or losers", we point out that change is a dynamic process. Next, we examine Phoenix, AZ and Detroit, MI in further detail.

Phoenix, Arizona is a "sun belt" city that was shaped by 20th century technological innovations that resulted in a demographic shift that has made it a boomtown. Founded in the 1880s, Phoenix was a small settlement in a desert that could not grow because of the lack of water. The relative accessibility and affordability of indoor air conditioning systems after World War II allowed people to consider Phoenix for year-round living rather than a winter escape for a few short months. The housing and settlement patterns in the 1950s favored automobile travel, suburban single-family housing typologies, and encouraged urban sprawl, see Figure 1.23 (US Census, n.d. and OSM). Government investments created a reliable, affordable water supply for the new city. In sum, technological innovations, laissez-faire capitalism, and voluntary migration encouraged new settlements to develop in areas that were previously not considered desirable, see Figure 1.24 (US Census, n.d.; NHGIS, n.d.), making Phoenix the fifth largest city in the United States. Despite its successes, Phoenix now struggles with the challenges of human-induced climate change, including hotter temperatures in summer, the challenges of maintaining a sustainable supply of water, and a growing homeless population that struggles to cope with rising costs of affordable housing.

Detroit, Michigan is a rust-belt city. In the first half of the twentieth century, Detroit grew in prominence as a result of the great migration of southern Blacks, as well as

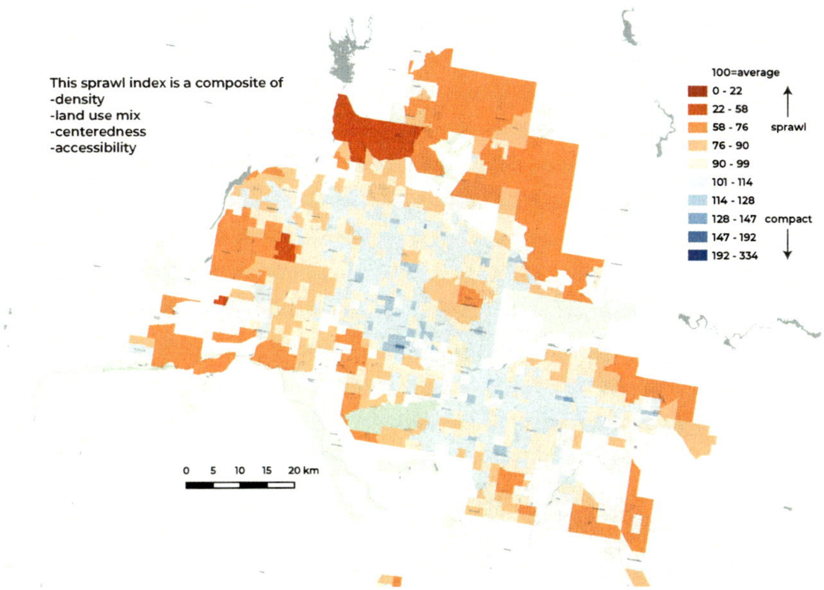


**FIGURE 1.23** Phoenix urban sprawl index

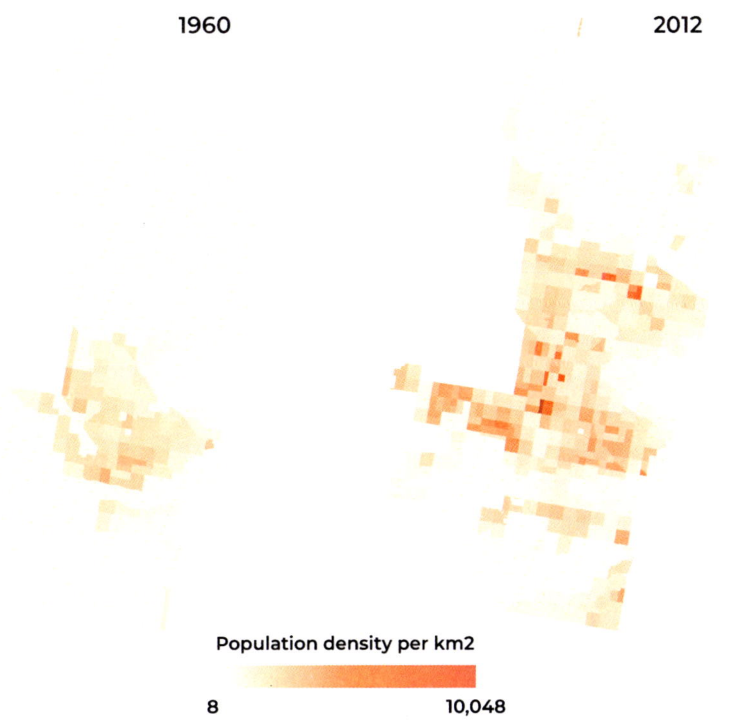


**FIGURE 1.24** Phoenix population growth, 1960–2012

immigrants from eastern and central Europe. The auto industry pioneered by Henry Ford established the principles of automation and mass production. The principles emphasized efficiency and also allowed low-skilled workers to become gainfully employed and part of a growing middle class. Workers in auto plants were able to have relatively clean and safe work and made a living wage, creating wealth across race and class lines. However, the auto industry's growth also contributed to de-densification and urban sprawl. As car culture evolved, cities like Detroit built networks of freeways to move people away from the city to residential suburbs, creating a host of negative consequences, chief among them being the destruction of thriving neighborhoods where Black people lived. Racial tensions caused urban riots, cementing segregation. The city of Detroit was crippled by white flight to the suburbs and the destruction of thriving Black neighborhoods through transportation policies that favored the car.

The current Metro-Detroit area map that includes Wayne, Macomb, and Oakland counties shows the stark contrast between inner city Detroit that shows serious population decline and areas of population growth in the outer suburbs. The effects of the decline of the automotive industry that began in the 1970s and 1980s have not been repaired. The consequences of population decline result in depressed home values, deterioration of housing stock, increase in number of vacant lots, and urban blight, see Figure 1.25 (US Census, n.d.; NHGIS, n.d.).

Historians, geographers, and urban scholars studying Detroit have documented the interlocking forces of private market decisions that privileged and advanced particular policy positions that were adopted by government planners and decision makers. In

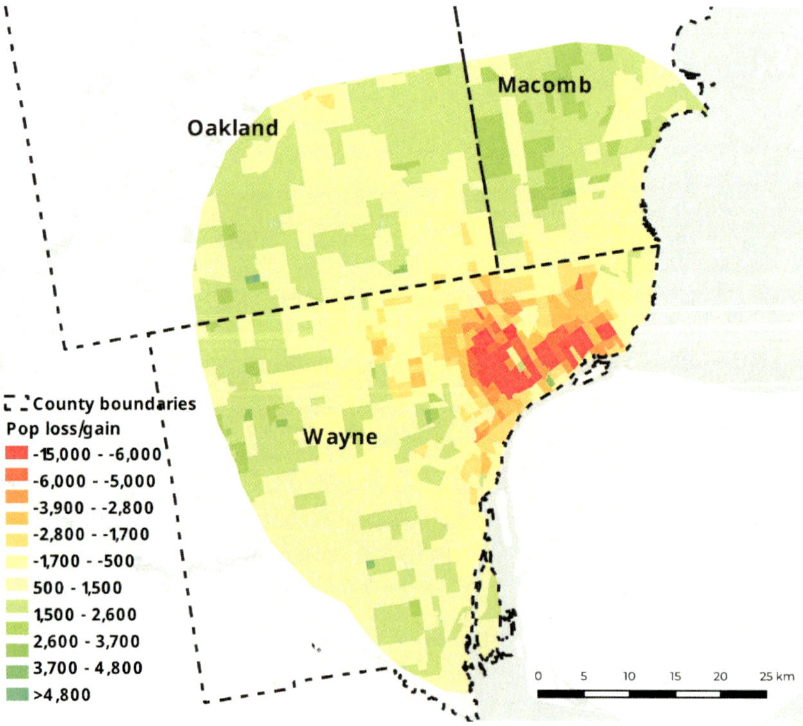


**FIGURE 1.25** Detroit population gain and loss

retrospect, the choices made in the 1950s have made it difficult to solve the depopulation crisis that remains a persistent challenge to present-day planners and city managers.

## 1.5 WHY SCALE MATTERS FOR HOUSING RESEARCH, PRACTICE, AND POLICY

Scale is one of the central anchors in geography, architecture, and spatial urban planning, yet it is a concept that is not well-understood by non-experts. Scale, in its everyday understanding, allows us to consider the relative size or complexity of an object, an event, or a process. Scale is also a useful concept to consider in representing real world objects, or processes on a map. Architects, for instance, can generate scaled drawings of their projects at a spatial scale of 1:10 or at a scale of 1:200. Each drawing serves a different purpose; for instance, a 1:4 scaled drawing may show the detail of an individual room, including the spatial relationship between the doors and windows of that room, whereas a drawing at 1:200 may better represent the building in relation to its site and setting.

Geographers and planners examining phenomena such as urbanization or environmental pollution represent processes. Processes are dynamic (change over time). Housing production, management, financing, and every other aspect of the housing enterprise occur across multiple spatial scales. Housing is inextricably linked to livability and quality of life. Thus, housing can and should be understood across different spatial scales, especially at the community and regional levels, rather than at a national level alone. It is only by understanding housing phenomena at the sub-regional and local levels can we understand geographic disparities in access to housing, for example, or assess whether national housing policies are having their intended effect in all regions, or whether one demographic group is being underserved or discriminated against in receiving financing to purchase homes. For example, housing starts (events) and urbanization (processes) are scale dependent and only understood by shifting/moving between spatial scales. While temporality also matters, understanding changes over space and across time requires consciously scaling up or down. Every aspect of housing can be examined at national, regional, and neighborhood levels. Using different scales to understand the same phenomenon can create a better understanding of the issue.

Another way to think about the value of thinking across spatial scales is to realize that humans experience the impacts of processes or events at different scales to arrive at different conclusions – for example, a walkable neighborhood may be appreciated at a local scale and be useful in creating a “walk score” (https://walkscore.com) for that neighborhood, while examining walkability at the scale of a city can be used to identify where new pedestrian paths must be laid to create equitable access. The only way to think across scales is to encourage spatial thinking, supported by available tools and methods. As a caveat, we note that because scale is linked to representation, there is a potential of manipulation/lying by adjusting the scale. Thus, a thorough understanding of scale is critical before deploying it to study housing phenomena (Albrecht, 2007; Ramasubramanian and Albrecht, 2018).

## 1.6 THE ROLE OF GIS IN ADDRESSING HOUSING CHALLENGES

This book’s primary focus is to understand and explain housing challenges in the United States using geographically referenced data and analyses. We are confident that our approaches can be used in different countries and cultural contexts by local experts who are familiar with the unique housing challenges in their country. The rest of the chapters in this book use a geographical lens to articulate our approach to examining housing challenges using Geographic Information Science (GIS) techniques. These techniques facilitate the (i) acquisition of data from diverse sources, (ii) specific analytical processes to query the data, and (iii) interesting ways to map and visualize results. Collectively, GIS assists in *communicating complex information* to diverse audiences, see Figure 1.26 from Perch Design Studio. We posit that GIS has not been effectively deployed in housing policy conversations at any scale and we hope that our contributions will be a step in the right direction.

As the authors of this book, we argue that this is an opportune time to use GIS to create new and compelling visual narratives that are anchored by data to understand housing in the context of neighborhood development. Housing (places of residence) seldom emerges/exists in isolation – where we live is closely related to where we work, socialize, shop, and worship. Examining housing as a single site (be it a single-family home or an affordable housing development) is pointless. Conversely examining housing policies such as whether a state policy supports the development of multi-family housing offers a simplistic and a non-spatial understanding of an inherently spatial phenomenon. The realtor's mantra, location, location, location, is relevant here. When realtors celebrate or talk up a location as a selling point, what they are trying to accomplish is to describe the non-monetary

**FIGURE 1.26** GIS as a communication device

value of the location relative to the neighborhood's amenities. Neighborhood matters! GIS enables end users to understand housing and neighborhoods.

In addition, GIS provides a working framework to allow experts and non-experts to collaborate easily and creatively. It is not about collecting data – different departments can remain the custodians of their own data. GIS makes it possible to combine data, to drive insights, and to change the way people make decisions. Most datasets, even those that were collected without any GIS in mind, have some locational reference. This is the unifying aspect about all GIS data that allows us to combine the data by location (see Chapter 4 for details) and visualize it in the form of a map. Geospatial visualizations increase engagement with internal and external stakeholders. Internally, GIS is used to answer questions such as "How are you making investments in communities, and what are the demographics of those communities?" Externally, GIS maps can demonstrate why the city is investing in certain areas and what progress they have made in achieving stated goals. GIS tools help create narratives that increase an understanding with all stakeholders. Envision Utah is one of many examples where GIS has been at the core of raising and responding to complex social policy questions.

One example for its ability to communicate complex housing policy information to diverse audiences is Envision Utah (https://envisionutah.org), a non-for-profit organization that aims to facilitate the rapid population growth of Utah in general and the northern part of the metro Salt Lake City region in particular. While in general, this is a good problem to have, the pains associated with such growth have to be addressed by planning efforts which are politically fraught as housing and transportation needs clash with environmental interests. Envision Utah uses GIS extensively to develop and discuss a range of regional growth scenarios ranging from car-oriented low-density to transit-oriented high-density alternatives with high levels of infill and redevelopment.

The visual nature of GIS enhances public outreach efforts. Envision Utah conducted over 30 public meetings and received input from some 3,500 online participants, in addition to the collaboration of over 60 stakeholder organizations. GIS provides the unique ability to be data-driven, while also visually communicating the consequences of one policy decision or the other. And whereas policy is usually equitable in its intent, the effects are often not because of different starting conditions at different locations. The spatial differentiation inherent in GIS inputs helps stakeholders to understand the pathways of a decision-making process in a complex context. As such, GIS serves both the planning expert and the proverbial Jane Q Public who does not want to be bothered with numbers but is presented with instantaneous cause and effects of tweaking one factor or the other.

## 1.7 OVERVIEW OF UPCOMING CHAPTERS

The book is intended to introduce contemporary housing issues to non-specialist audiences and to encourage housing policy professionals and housing experts to use GIS concepts, methods, and techniques to investigate housing-related policy and implementation questions. As authors, we are clear that the context and the application domain (housing) and the questions posed to understand, explain, and shape housing policy must determine the use of the methods – in this case, the use of GIS mapping and spatial analysis. Therefore, we begin by framing the first chapter in contextualizing housing in the United States. In Chapter 1, we propose that housing, understood to be a basic human need in all societies, is much more than the provision of shelter from the elements. Housing is a cultural phenomenon, in that housing morphologies and settlement patterns are as likely to reflect a society's dominant cultural ethos which may be as significant or more significant than building with considerations about suitability of materials, methods of construction, or costs. American urbanization and housing settlement patterns were influenced by the City Beautiful movement and societal considerations that sought to reduce density and overcrowding in urban areas. The chapter proposes that the application of a geographical lens can create a robust understanding of housing related issues at the neighborhood/human scale and advocates for the use of Geographic Information Science concepts, methods, and techniques to formalize geographical analyses of housing questions. Maps and data are used to explain housing geographies and highlight how the use of geographically referenced, publicly available information can be used to support policymaking.

Chapter 2 addresses the demographic shifts in the United States since the early 20th century to establish that demographic realities, regardless of their cause, influence housing production. At the same time, housing production innovations were made possible because of technological advances. Just as in the late 19th century, indoor plumbing and sanitary sewer systems allowed changes in the layout of individual houses and apartment buildings, the invention of indoor air conditioning systems allowed large-scale settlements to develop in previously inhospitable climatic zones. Likewise, advances in refrigerated trucking, and the development and growth of an automobile-oriented culture influenced American urbanization since the 1920s, trends that accelerated after 1945 as a result of the post war baby boom. Suburbia was "invented" to advance an ideal of a nuclear family (with a working father and a stay-at-home mom), who lived in "safe" suburban residential environments spatially separated from unhealthy and unsafe urban workplace environments. This cultural ideal has prevailed for over 60 years and shows no signs of abating. Land use policies and practices supported and advanced these cultural ideals. Zoning imposed a pseudo-order on the landscape and was established using racist and exclusionary practices that created segregated neighborhoods. The chapter also discusses the rise and decline in investments to create public housing and ends with a brief discussion of gentrification.

Chapter 3 expands the readers' understanding about housing typologies. Non-specialists hearing the word "housing" typically use their own personal understandings of housing establishing a rudimentary binary classification that distinguishes between owner-occupied housing and renter-occupied housing. While this is an important distinction, there are additional architectural distinctions that become significant in the production and management of housing. The range of housing typologies discussed in this chapter creates distinctions by function: (i) single-family housing, (ii) multi-family housing, and (iii) institutional living quarters. From a planning and design perspective, each of these types of housing typologies can be further broken down based on sub-categories such as (i) architectural styles (e.g., a single-family detached house), allowable height/volume (e.g., a non-elevator, walk-up building in a residential zone), number of individuals or households accommodated (e.g., group quarters such as college dorms), and ownership (e.g., condominiums). The chapter also discusses newer physical planning innovations in the housing sector such as the use of manufactured homes to address the housing shortage among low-wage workers, the legal and illegal conversions of homes to add space for expanding families such as a mother-in-law unit, or a rental unit to generate income for the house owner. The chapter also discusses policy innovations such as the Rental Assistance Demonstration (RAD) program developed by the US Department of Housing and Urban Development (HUD) that attempts to preserve affordable housing. The chapter concludes with a discussion about geographical data and the use of geospatial indicators that can be used to understand the land use planning implications associated with housing. Housing specialists eager to learn about the use of GIS to support their work should begin with a careful reading of Section 3.5.

Chapter 4 shifts the focus from housing issues to a consideration of GIS. Geographic Information Science concepts require a preliminary understanding of data sources, databases, database organization principles, and data quality. The chapter begins with a discussion of different types of data sources. The primary data source for housing research comes from the US Census. Additionally, historical census data has been compiled and organized in formats suitable for research and analysis through the National Historical GIS. In order to take advantage of GIS, data needs to be locationally referenced, creating a unique spatial identifier – such as a street address to which other information can be linked. Chapter 4 also discusses how to create and use derived housing variables (that are computed or estimated), and the challenges of working in data poor environments. A discussion of data quality includes the Modifiable Areal Unit Problem (MAUP), which is the cause of ecological fallacies in analyzing and reporting GIS data. The chapter concludes with a reminder to readers to be creative in identifying non-conventional data sources and engaging citizens in conducting housing research to solve the problem of "lack of data" at the neighborhood level. The chapter does not claim to be an introduction to GIS (whole textbooks have been written about how to use GIS) but it gets housing specialists prepared to have conversations and collaborations with GIS professionals.

Chapter 5 builds on the previous chapter and prepares housing researchers to understand the principles of spatial coordinate systems that allow geo-referenced data points to be displayed correctly on a map. It also explains the geo-relational principle wherein a unique location reference in one database or data table is linked to the same unique reference in another dataset. Using a locational reference to link different sets of data lies at the basis of GIS. Data *about* a location (in GIS terms, called attributes) gathered from different sources can be linked and analyzed. For planners and housing researchers, this means that population data about a neighborhood (ZIP code) can be linked to housing prices in the same area and can also be linked to education characteristics at the same location. The chapter continues with a discussion of basic GIS operations that are used in GIS-based analyses and touches upon advanced methods such as spatial regression.

Chapter 6 is the most technical chapter in this book. GIS novices are urged to use a companion GIS textbook such as Albrecht (2007) and a planning methods book e.g., Ramasubramanian and Albrecht (2018) to extract value from this chapter. Chapter 6 moves from the realm of using GIS to understand housing issues to considering the complex policy questions that preoccupy housing specialists and urban planners. Progressive housing activists and policymakers are consumed by addressing the housing supply crisis discussed in Chapter 1 by increasing residential densities. The authors discuss how GIS can support framing these policy conversations. Each section tackles complex challenges where multiple layers of data and evidence are linked to produce a coherent narrative to advocate for a particular set of policies. In this chapter, GIS maps are recognized as services that create just-in-time analyses for end users. GIS is also more robustly integrated with 3 D modeling and visualization, requiring advanced technical skills. The latest investments in GIS for housing use digital twins – where digital models of the urban environment are created in great detail to allow the testing of different scenarios or options. These technological advances push the boundaries of what is possible using GIS, and support future-oriented planning and decision making.

Throughout this book, we have discussed the power and potential of using a geographical lens to examine housing issues at different spatial scales. In Chapter 7, which concludes this book, we remind readers that GI technologies and applications facilitate academic inquiry but more importantly allow for a range of stakeholders to examine housing questions in relation to other city development challenges such as addressing infrastructure or transportation needs. Since housing is central to the lives of everyday people and housing challenges are experienced at the neighborhood scale, we have argued about the value and need for housing analyses to be conducted and communicated at the neighborhood/sub-city scale. We encourage educators in the design and planning professions to integrate policy and planning conversations – to further encourage professionals working in the built environment sector to work collaboratively to address housing production/supply challenges. We encourage bold thinking and forward-looking solutions to address the enduring housing crises in America to create sustainable and humane living alternatives for future generations.

## REFERENCES

AARP Research, 2018. *2018 Home and Community Preferences: A National Survey of Adults Ages 18-Plus*. Washington, DC: American Association of Retired Persons. https://www.aarp.org/content/dam/aarp/research/surveys_statistics/liv-com/2018/home-community-preferences-survey. doi:10.26419-2Fres.00231.001.pdf, last accessed 6 March 2023.

Albrecht, J, 2007. *Key Concepts and Techniques in GIS*. London: Sage Publications.

Beauregard, R A, 2006. *When America Became Suburban*. Minneapolis: University of Minnesota Press.

BLS, 2022. *Consumer Expenditures for the San Francisco Metropolitan Area, 2020–2021*. Washington, DC: Bureau of Labor Statistics. https://www.bls.gov/regions/west/news-release/consumerexpenditures_sanfrancisco.htm#:~:text=Housing%20accounted%20for%2041.9%20percent,higher%20than%20the%20national%20average, last accessed 6 March 2023.

Burchell, R, Downs, A, McCann, B, and Mukheerjee, S, 2005. *Sprawl Costs: Economic Costs of Unchecked Development*. Washington, DC: Island Press.

Carbone, J, Clift, J, Wyllie, T, and Smyth, A, 2022. "Housing Unit Type and Perceived Social Isolation Among Senior Housing Community Residents". *The Gerontologist*, 62(6): 889–899.

Caro, R, 1974. *The Power Broker: Robert Moses and the Fall of New York*. New York: Vintage Books.

Cavanaugh, E, 2020. *The Routledge Handbook of the History of Settler Colonialism*. Philadelphia, PA: Routledge.

Chey, K, 2017. *Multi-Unit Housing in Urban Cities: From 1800 to Present Day*. Philadelphia, PA: Routledge.

Congressional Budget Office, 2022. *The Demographic Outlook: 2022–2052*. https://www.cbo.gov/system/files/2022-07/57975-demographic-outlook.pdf, last accessed 2 July 2023.

Congressional Research Service, 2023. *An Introduction to the Low Income Housing Tax Credit. RS22389*. Washington, DC: Congressional Research Service. https://crsreports.congress.gov/, last accessed 30 April 2023.

Desmond, M, 2017. *Evicted: Poverty and Profit in the American City*. New York: Crown.

Doughterty, C, and Casselman, B, 2022. "We Need to Keep Building Houses Even If No One Wants to Buy". *New York Times*, July 23. https://www.nytimes.com/2022/07/23/business/housing-market-crisis-supply.html, last accessed 6 March 2023.

Fallon, K, 2021. *Naming Housing as a Human Right Is the First Step to Solving the Housing Crisis, Housing Matters*. Washington, DC: Urban Institute. https://housingmatters.urban.org/articles/naming-housing-human-right-first-step-solving-housing-crisis, last accessed 6 March 2023.

Fisher, C, 1962. "Land Use Control Through Zoning: The San Francisco Experience". *Hastings Law Journal*, 13: 3. https://repository.uchastings.edu/hastings_law_journal/vol13/iss3/3, last accessed 6 March 2023.

Fitzpatrick, K, and Spialek, M, 2020. *Hurricane Harvey's Aftermath: Place, Race, and Inequality in Disaster Recovery*. New York: NYU Press.

Forsyth, A, 2012. "Defining Suburbs". *Journal of Planning Literature*, 27(3): 270–281. doi:10.1177/0885412212448101.

Foundation for Community Association Research, 2017. 2016 *Community Association Fact Book Part Six: Comparison of Condominium and Non-Condominium Residents Age 55 and Over*. Falls Church, VA: Foundation for Community Association Research. https://foundation.caionline.org/, last accessed 2 July 2023.

Frey, W H, 2019. *America Is Not Full. Its Future Rests With Young Immigrants*. https://www.brookings.edu/articles/america-is-not-full-its-future-rests-with-young-immigrants/, last accessed 2 July 2023.

Galster, G, 2008. "United States Housing Scholarship since 1968". *Journal of the American Planning Association*, 74: 1. doi:10.1080/01944360701792231

Gans, H, 1967, 2017. *The Levittowners: Ways of Life and Politics in a New Suburban Community*. New York: Columbia University Press.

Goodman, P, 2022. "A Normal Supply Chain? Its 'Unlikely'". *New York Times*, February 6. https://www.nytimes.com/2022/02/01/business/supply-chain-disruption.html, last accessed March 6, 2023.

Grant, J, Mottet, L, Tanis, J, Harrison, J, Herman, J, and Keisling, M, 2011. *Injustice at Every Turn: A Report of the National Transgender Discrimination Survey*. Washington, DC: National Center for Transgender Equality and National Gay and Lesbian Task Force.

Hall, P, 2014. *Cities of Tomorrow: An Intellectual History of Urban Planning and Design since 1880*. West Sussex: Wiley Blackwell.

Haub, C, and Sharma, O, 2018. *How People in India "Really" Live*. Population Reference Bureau Report. https://www.prb.org/resources/how-people-in-india-really-live/, last accessed 6 March 2023.

Hayden, D, and Wark, J, 2005. *A Field Guide to Sprawl*. New York: Norton.

Healy, J, 2016. "Underwater in the Las Vegas Desert, Years after the Housing Crash". *New York Times*, August 2. https://www.nytimes.com/2016/08/03/us/las-vegas-2008-housing-crash.html, last accessed 6 March 2023.

Henry, M, de Sousa, T, Roddey, C, Gayen, S, and Bednar, T, 2021. *The 2020 Annual Homeless Assessment Report (AHAR) to Congress, Part 1: Point-In-Time Estimates of Homelessness*. https://www.huduser.gov/portal/sites/default/files/pdf/2020-AHAR-Part-1.pdf, last accessed 6 March 2023.

Hixon, W, 2013. *American Settler Colonialism: A History*. New York: Palgrave Macmillan.

Howard, E, 1898. *Tomorrow: A Peaceful Path to Real Reform*. London: Swan Sonnenschein & Co., Ltd.

Howard, E, 1902 (reissued 1965). *Garden Cities of Tomorrow*. Cambridge, MS: MIT Press.

Hoyt, H, 1939. *The Structure and Growth of Residential Neighborhoods in American Cities*. Washington, DC: Federal Housing Administration.

Jackson, K, 1985. *Crabgrass Frontier: The Suburbanization of the United States*. London: Oxford University Press.

Jayapal, P, and Meng, G, 2021. *Housing is a Human Right Act of 2021*. https://jayapal.house.gov/wp-content/uploads/2021/06/Housing-is-a-Human-Right_Jayapal.pdf, last accessed 6 March 2023.

Jones, K, Getter, D, and Scott, A, 2017. *An Overview of the Housing Finance System in the United States*. Congressional Research Service Report, R 42995. https://crsreports.congress.gov/, last accessed 6 March 2023.

Kaysen, R, 2023. "A Housing Market Hangover". *New York Times*, January 3. https://www.nytimes.com/2022/12/30/realestate/housing-market-prices-interest-rates.html, last accessed 6 March 2023.

Kelly, B, 1993. *Expanding the American Dream: Building and Rebuilding Levittown*. Albany, NY: State University of New York Press.

Koop, A, 2022. *Ranked: The 25 Countries receiving the most US Immigration Visas*. https://www.visualcapitalist.com/countries-receiving-most-us-immigration-visas/, last accessed 1 June 2023.

Kunstler, J, 1994. *The Geography of Nowhere: The rise and decline of America's Man-Made Landscape*. New York: The Free Press.

Lawton, M, and Nahemow, L, 1973. "Ecology and the Aging Process". In Eisdorfer, C, and Lawton, M (Eds.), *The Psychology of Adult Development and Aging*, pp. 619–674. Washington, DC: American Psychological Association. doi:10.1037/10044-020.

Lee, B, Tyler, K, and Wright, J, 2010. "The New Homelessness Revisited". *Annual Review of Sociology*, 36(1): 501–521.

LeGates, R T, and Stout, F (Eds.), 2019. *The City Reader*, 6th edition. Philadelphia, PA: Routledge.

Levittown, P A. Wikimedia Commons, n.d., https://commons.wikimedia.org/wiki/File:LevittownPA.jpg.

Ley, D, and Tuchener, J, 1999. "Immigration and Metropolitan House Prices in Canada". *Research on Immigration and Integration in the Metropolis: Working Paper #99-09*. Vancouver, BC: Vancouver Center of Excellence.

Madden, D, and Marcuse, P, 2016. *In Defense of Housing: The Politics of Crisis*. New York: Verso.

Means, R, 2007. "Safe as Houses? Ageing in Place and Vulnerable Older People in the UK". *Social Policy & Administration*, 41(1): 65–85. doi:10.1111/j.1467-9515.2007.00539.x.

Meyer, B, Wyse, A, and Corinth, K, 2022. *The Size and Census Coverage of the U.S. Homeless Population: Working Paper No.* 2022-78. Chicago, IL: Becker Friedman Institute for Economics at the University of Chicago. https://bfi.uchicago.edu/working-paper/the-size-and-census-coverage-of-the-u-s-homeless-population/, last accessed 6 March 2023.

Muller, T, and Espenshade, T, 1985. *The Fourth Wave: California's Newest Immigrants*. Washington, DC: The Urban Institute Press.

Myers, D, and Pitkin, J, 2013. *Immigrant Contributions to Housing Demand in the United States: A comparison of Recent Decades and Projections to 2020 for the States and the Nation*. https://papers.ssrn.com/sol3/papers.cfm?abstract_id=2239077, last accessed 2 June 2023.

National Alliance to End Homelessness, 2022. *State of Homelessness Report*, 2022. https://endhomelessness.org/homelessness-in-america/homelessness-statistics/state-of-homelessness/, last accessed 6 March 2023.

National Coalition for the Homeless, 2009. "Mental Illness and Homelessness". *Fact Sheet*, July 2009. https://www.nationalhomeless.org/factsheets/Mental_Illness.html, last accessed 6 March 2023.

National Rural Housing Coalition (NRHC), 2014. *Federal Strategies to Preserve Access to Affordable Rental Housing in Rural Communities*. Washington, DC: NRHC

NHGIS, n.d. *Generic historical US Census data archived by National Historical GIS*. Online resource, available at nhgis.org.

Office of the United Nations High Commissioner of Human Rights (OHCHR), 2009. "The Human Right to Adequate Housing". *Fact Sheet*, 21. https://www.refworld.org/docid/479477400.html, last accessed 6 March 2023.

Oldenburg, R. 1989. *The Great Good Place: Cafes, Coffee Shops, Community Centers, Beauty Parlors, General Stores, Bars, Hangouts, and How They Get You Through the Day*. New York: Paragon House.

Olivet, J, 2022. *Collaborate, Don't Criminalize: How Communities Can Effectively and Humanely Address Homelessness*. Washington, DC: U.S. Interagency Council on Homelessness. https://www.usich.gov/news/collaborate-dont-criminalize-how-communities-can-effectively-and-humanely-address-homelessness/, last accessed 30 May 2023.

Park, R, Burgess, E, and McKenzie, R, 1925 (reissued 1967). *The City*. Chicago, IL: The University of Chicago Press.

Parolek, D, 2020. *Missing Middle Housing: Thinking Big and Building Small to Respond to Today's Housing Crisis*. Washington, DC: New York Island Press.

Perch Design Studio. *Interactive Map Table*. https://perchmade.com/experience/interactive-map-table/, last accessed 13 July 2023.

Perry, C, 1929. "The Neighborhood Unit". In LeGates, R, and Stout, F (Eds.), *The City Reader*, 5th edition, pp. 488–498. Philadelphia, PA: Routledge.

Ramasubramanian, L, 2010. *Geographic Information Science and Public Participation*. Heidelberg, Germany: Springer.

Ramasubramanian, L, and Albrecht, J, 2018. *Essential Methods for Planning Practitioners: Skills and Techniques for Data Analysis, Visualization, and Communication*. New York: Springer.

Rapoport, A, 1969. *House Form and Culture*. New York: Pearson.

Rothstein, A, 1938. "Slums, Pittsburgh, Pennsylvania [Graphic]: 1938 July (1 Negative: Nitrate; 35 mm)". *The Hill*. Library of Congress, Prints and Photographs Division. https://lccn.loc.gov/2017723699.

Rothstein, R, 2018. *The Color of Law: A Forgotten History of How Our Government Segregated America*. New York: Norton Publishing.

Saiz, A, 2007. "Immigration and Housing Rents in American Cities". *Journal of Urban Economics*, 61(2): 345–371. doi:10.1016/j.jue.2006.07.004.

Schön, D, and Rein, M, 1994. *Frame Reflection: Toward the Resolution of Intractable Policy Controversies*. New York: Basic Books.

Sisson, P, 2022. "For Want of an Insulation Screw". *New York Times*, November 22. https://www.nytimes.com/2022/11/15/business/construction-costs.html, last accessed 6 March 2023.

Southworth, M, and Joseph, E B, 2003. *Streets and the Shaping of Towns and Cities*. Washington, DC: Island Press.

Sugiyama, M, Chau, H, Takumi, A, Yusuke K, Jamei, E, Veeorja, P, Mori, K, and Sugiyama, T, 2023. "Third Places for Older Adults Social Engagement: A Scoping Review and Research Agenda". *Gerontology*, 63: 1149–1161, doi:10.1093/geront/gnac180.

US Census, n.d. *Generic American Community Survey (ACS) data*. Online resource, available at census.gov/programs-surveys/acs/.

U.S. Census Bureau, 2014. *Current Population Survey: March and Annual Social and Economic Supplements*. Washington, DC: U.S. Census Bureau.

U.S. Census Bureau, 2021. *American Community Survey, 2021: American Community Survey 5 Year Estimates*. Washington, DC: U.S. Census Bureau.

US Census, 2021b. *Historical Population Change Data (1910–2020)*. https://www.census.gov/data/tables/time-series/dec/popchange-data-text.html, last accessed 24 May 2023.

USDA, 2022. *US Department of Agriculture*. Online resource available at sc.egov.usda.gov/data/files/MFH/, last accessed 14 April 2023.

USHUD, 2022. *Point-in-Time Count and Housing Inventory Count, Homelessness Data Exchange (HDX)*, Washington, DC: U.S. Department of Housing and Urban Development. Online resource. https://www.hudexchange.info/programs/hdx/pit-hic/, last accessed 12 February 2023.

United Nations General Assembly (UNGA), 1948. *The Universal Declaration of Human Rights, UDHR*. New York: United Nations General Assembly. https://www.un.org/en/about-us/universal-declaration-of-human-rights, last accessed 6 March 2023.

Walmsley, D, and Lewis, G, 1993. *People and Environment: Behavioral Approaches in Human Geography*, vol. 2. Philadelphia, PA: Routledge.

# 2 Social, Demographic, and Technological Shifts and Their Impacts on Housing

## 2.1 20TH CENTURY DEMOGRAPHIC SHIFTS

According to the US Census, the population of the United States in 1901 was a little under 78 million people, and throughout the 20th century, the population grew gradually, at an average rate of growth of between 1% and 2% every year (US Census, 2000, 2021a). There were some years when the growth rate declined, for instance during the war years, but in general the US population has continued to grow in overall numbers. In 2020, the population of the United States was recorded as "331,449,281 as of April 1, 2020, an increase of 7.4% since the 2010 Census" (US Census, 2021b). Demographers further explore the composition of the population, in terms of age and gender, consider birth and death rates, and track different factors that can explain population fluctuations. In Chapter 1, Section 1.4 we noted that housing is directly and indirectly affected by demographic shifts. The first factor is how many new people are born – that is related to the fertility rate. The fertility rate in the United States has been declining since 1960. However, immigration has bolstered population growth. Immigration fueled growth has been shaped by American foreign policy and immigration policy, thereby impacting the composition and household structure of foreign-born populations (Pew Research Center, 2015). For example, the change in American immigration policy after the passage of Immigration and Nationality Act of 1965 is credited with the rapid growth of Asian populations who were artificially prevented from entering the United States legally during the previous decade (Lee, 2016). The actual numbers, the population characteristics, and the motivation for immigration influenced regional variations in settlement patterns as well as the type of housing that was needed, see Figure 2.1 (data source, US Census, 2021a).

### 2.1.1 Settlement Patterns and Segregation

Settlement patterns in the United States in the 20th century co-evolved alongside and because of technological and political shifts. The transformations were non-linear and violent. By 1900, the United States was already making a shift from a largely rural and agrarian society to an early industrial society supported by waves of immigrants. In the late 1800s, most of the immigrants arrived in New York and traveled

DOI: 10.1201/9780429243066-2

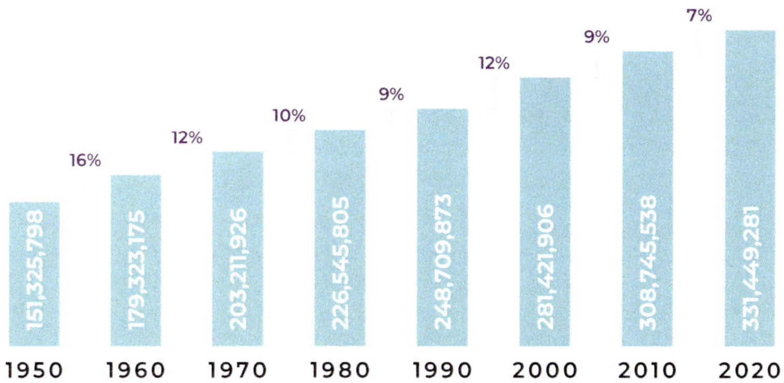


**FIGURE 2.1** US population change from 1950 to 2020

to other cities along the eastern seaboard and further inland to Chicago and points west. Figure 2.2 (US Census, n.d.; NHGIS, n.d.) shows on a state-by-state basis, during which year over the past 150 years each state had its highest percentage of the total US population. Growth and expansion were encouraged and endorsed by the government. The government also created laws severely repressing Blacks, putting in place the early frameworks of segregated settlement patterns (Cavanaugh, 2020) Figure 2.3 (US Census, n.d.; NHGIS, n.d.) shows the date in which each state crossed the threshold to majority urban.

### 2.1.2 The Great Migration

Both in the industrial North and the agricultural South, segregation was a persistent challenge in the late 1800s. Southern Blacks began migrating to northern cities like Detroit and Chicago in search of work in factories, experiencing two major push factors: (i) the lack of viable economic opportunities in farming and (ii) the climate of fear caused by the violent actions of hate groups like the Klan. Black migrants to the North found work, although that work was often dirty and dangerous. Segregation of African Americans was sanctioned by law and many cities passed laws that actively discriminated against Blacks. In the larger cities, new immigrants found themselves clustered into ethnic enclaves because of discriminatory housing policies. Even when they did not have to confront racist laws, they experienced de-facto segregation because of limited access to housing, resulting in over-crowded living conditions in many American cities including New York City and Chicago. In NYC, Blacks settled in Harlem which became a city within a city – the cultural and economic heart of a Black metropolis – and in Chicago around Hyde Park (Wilkerson, 2011).

The population of the United Students has grown steadily, and Figure 2.4 (data source, US Census, 2021a) exemplifies how the Hispanic population has grown since the 1980s. Figure 2.5 (data source, US Census, 2021a) further identifies how different states absorbed this growth, with California, Texas, Florida (states along

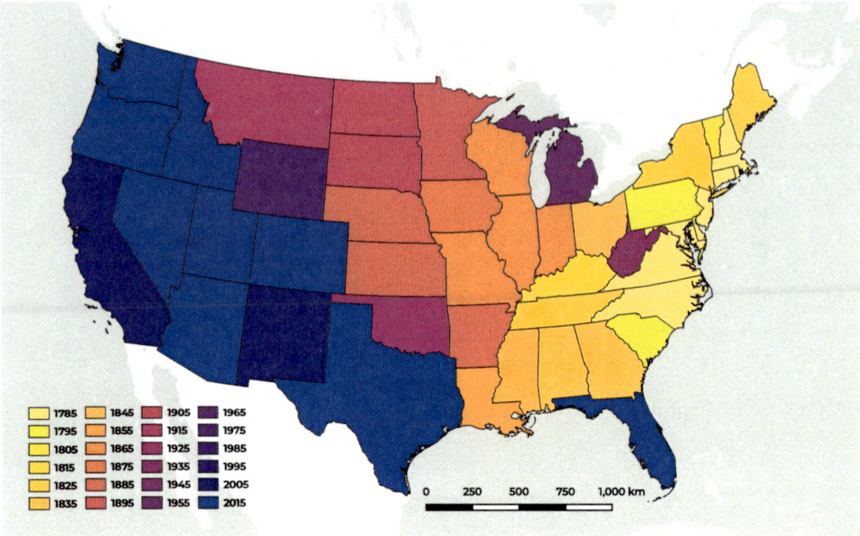


**FIGURE 2.2** US states heyday

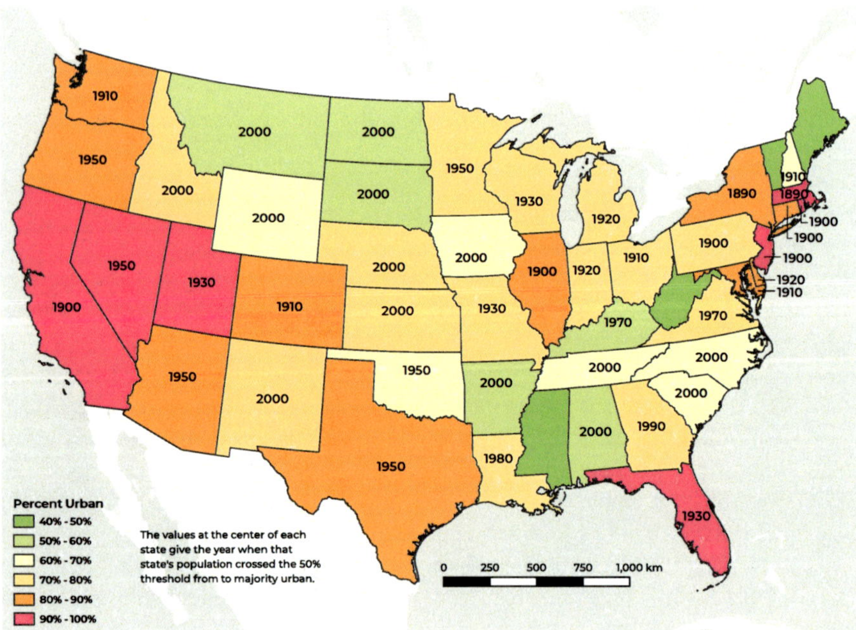


**FIGURE 2.3** US percent urban

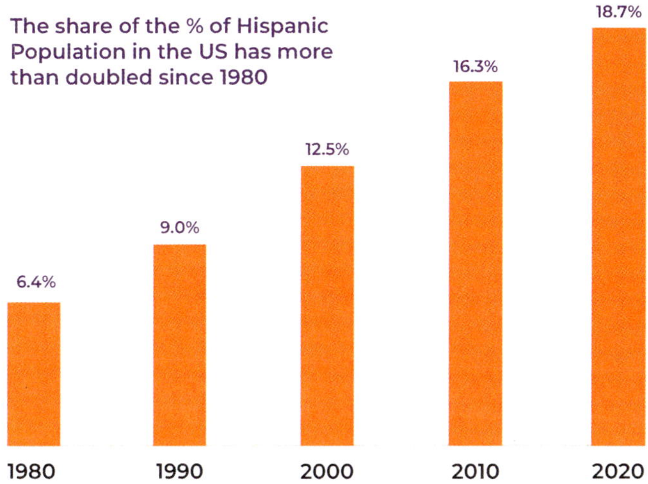


FIGURE 2.4 Percent of Hispanic population 1980–2020

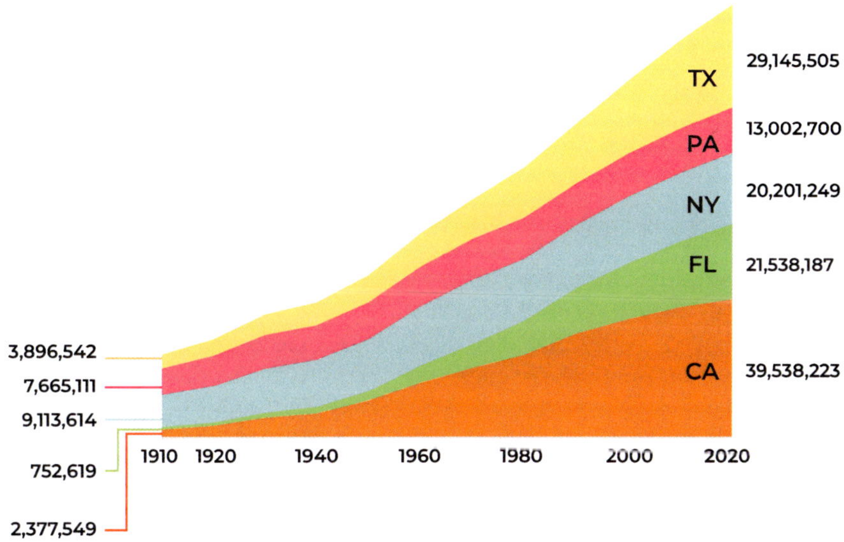


FIGURE 2.5 Population change from 1910 to 2020 of the five most populous states

the country's southern border) and the three most populous states in 2020 with New York and Pennsylvania rounding out the top five in terms of total population. However, it is useful to note that the numbers in New York have been shrinking since 2016 and Pennsylvania's population has stopped growing since 2019. The growth in the Hispanic (Latino) population has influenced the overall population growth. Figure 2.4 shows that the percentage of the Hispanic population as a share of the total population grew from 6.4% in 1980 to 18.7% in 2020. The Hispanic population is

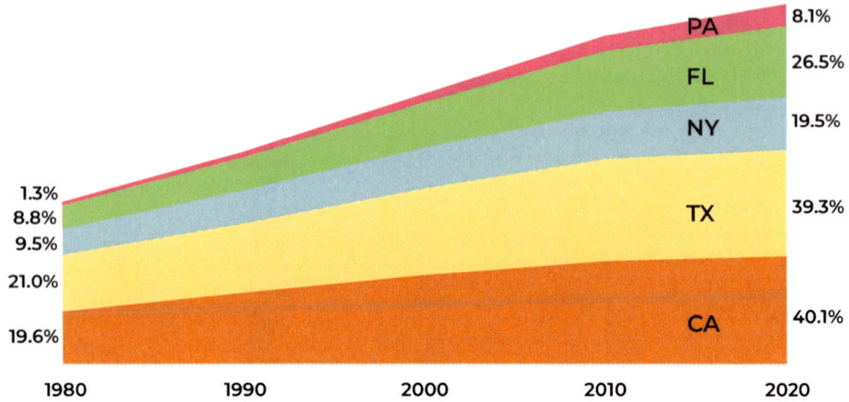


**FIGURE 2.6** Percent Hispanic change from 1980 to 2020 of the five most populous states

projected to grow further in the coming decade. One of the challenges in considering the Hispanic population is that the category "Hispanic" is a cultural/ethnic category that can be interpreted differently and cannot be easily combined with existing racial categories such as White/Caucasian and Black/African American. In states such as California with high Hispanic populations, as seen in Figure 2.6 (data source, US Census, 2021a), these distinctions can become blurred. In both Texas and California, Hispanics comprise nearly 40% of the population, and their choices and preferences are likely to directly influence housing and urban development.

The Census has changed the way it asks questions about race over decades. In order to make our visualizations easy to understand, we made the decision to examine data in two categories, White and Non-White. Figure 2.7 (NHGIS) shows the movement of non-White populations in a series of six county-level maps for the years 1900, 1920, 1950, 1970, 2000, and 2020. Non-White populations were always present throughout the United States although these populations were more concentrated in the south and southeast between 1900 and 1920. The post-war years, 1950 and 1970, show more dispersion and a movement westward. The 2020 map offers the clearest evidence that other than counties in rural areas, Non-White populations are found in every county in the United States. Figure 2.8 (data source, US Census, 2021a) quantifies the percentage of Non-White populations which grew from 10.5% of the total population in 1950 to 38.4% in 2020.

### 2.1.3 Public Housing

Housing those who do not have the private means to develop and house themselves either through home ownership or through the private rental market has been a challenge for government policymakers in the United States throughout the last century. According to Vale and Freemark (2012, p. 382),

> American public housing is a) a 25-year series of efforts to accommodate the upwardly mobile working class between 1935 and 1960, and thereafter the worthy elderly; b) a 30-year consolidation of the poorest into welfare housing between 1960 and the mid 1980s, coupled by efforts to introduce direct private-sector involvement in public

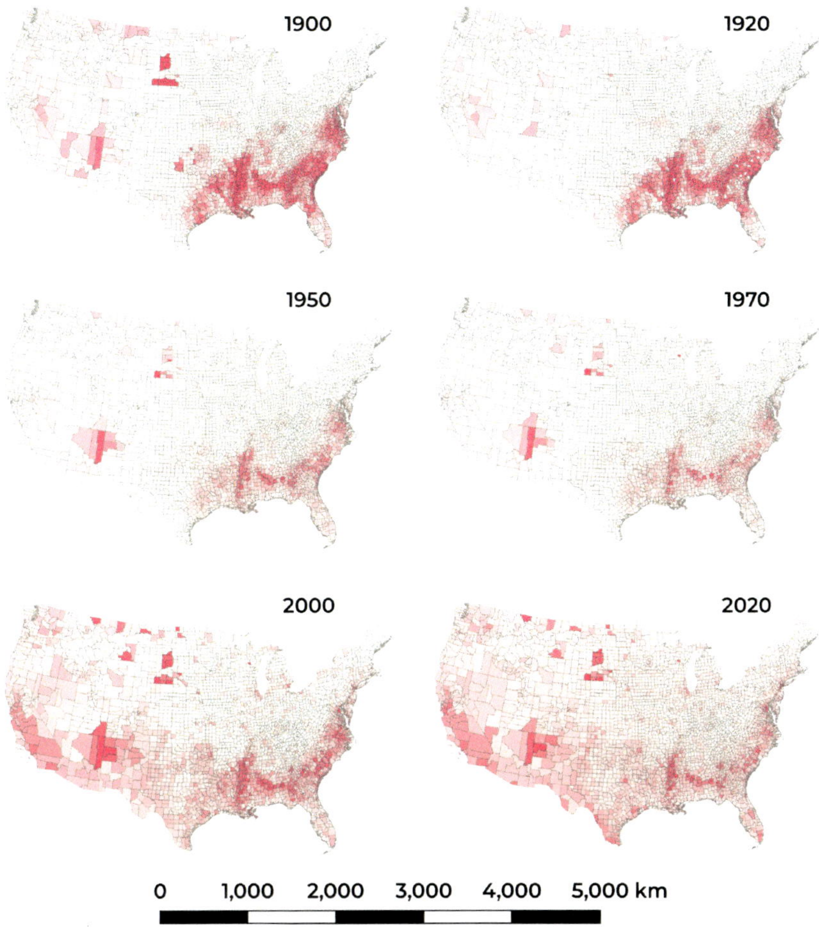


**FIGURE 2.7** Settlement patterns by race. Non-White populations across the United States over time

> housing and other programs; and c) a series of programs and policies since the mid-1980s to return more of public housing to a less-poor constituency, while furthering growth in other kinds of both deep and shallow subsidy programs through mixed-finance projects and tax-code intervention.

When mention is made of public housing, most Americans immediately think about "the projects", high-rise apartment complexes in big cities like New York and Chicago. The negative connotations about public housing relate to real and stereotypical concerns about crime, safety, and social disorder. The planned demolition of the Pruitt Igoe complex of public housing developments in St. Louis symbolized the governmental and societal disenchantment with public housing. Pruitt Igoe, built in the 1950s, was torn down in the early 1970s. It is important to note that

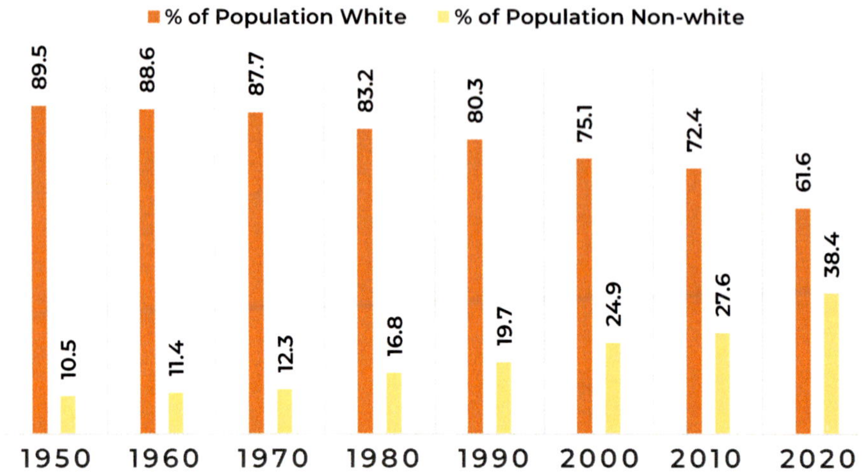


**FIGURE 2.8** US population: share of minority populations from 1950 to 2020

the production, supply, and even the location of public housing are closely linked with national values about how to serve disadvantaged populations using public dollars. Thus, public housing developments were austere, imposed social sanctions about "appropriate behavior", concentrated poverty by being situated in low-income neighborhoods, were allowed to deteriorate through a pattern of deferred maintenance, and prevented individual agency by rigid adherence to formal rules (Bloom et al., 2015; Schwartz, 2021).

Outside the United States, Vienna's "Social Housing" is known for its high quality and affordable housing for low-income residents, as well as its innovative urban design and architecture. The city has a long history of investing in social housing, dating back to the early 20th century, and has become a model for other cities around the world. More than 60% of the city's 1.8 million residents live in social housing. Additionally, the city places a strong emphasis on sustainability, with many of the housing developments featuring green spaces and energy-efficient design elements (Holzner and Huberman, 2022).

Even by European standards, Vienna is an outlier in that its municipal government has continually made the preservation of its social housing stock a central aspect of its political identity at the same times as other European cities privatized social housing in the 1980s and 1990s. Buildings built nearly a century ago continue to provide comfortable and well-maintained housing for the city's residents. Vienna's other housing policies, such as rent control,[1] undergird the city's ability to maintain and expand its social housing stock. In sharp contrast to the United States, public attitudes towards subsidized housing are very different (it helps that a majority of Viennese benefit from these subsidies). Housing is seen as a public benefit rather than as alms for the poor with the city spending about 11% of its municipal budget on social housing (Holzner and Huberman, 2022). In consequence, the social housing estates are full of middle-class amenities that are cherished by the tenants and imbue pride and ownership that are very different from what tenants in US projects experience. The individual economic security afforded by social housing results in active participation on local housing councils that support comprehensive neighborhood development from car-free streets to kindergartens and social clubs, which in turn help to keep crime rates low. While the city of Vienna has a formidable GIS program with an impressive amount of open-source data that advances evidence-based decisions, Vienna's successes are also a result of a commitment to keep housing generally affordable that spans left-right party ideologies. In addition, the high degree of community participation ensures that diverse perspectives influence decision-making in all sectors related to community quality-of-life, i.e., beyond a narrow focus on affordable housing policy.

The "heat map" in Figure 2.9 (USHUD, 2023) depicting concentrations of people in public housing shows robust concentrations in the Bos-Wash corridor, in Pittsburgh, Cincinnati, and Chicago a little further to the west, and in Raleigh, Memphis, and Birmingham in the South. The United States never had a robust supply of public housing units, when compared with the population's needs. Furthermore, there has been a steady decline in the production and availability of public housing units over time resulting in fewer numbers of people in public housing (see Figure 2.10 (Office of Policy Debelopment and Research. *HUD User*, n.d.)).

## 2.2 TECHNOLOGICAL SHIFTS

Natural and human-induced disasters prompted city planners to alter building form and implement zoning regulations in an effort to uphold societal ideals by zoning for light and air to create better living conditions for the urban poor who lived in close quarters in squalid settlements (tenements). The availability of new materials, new methods of construction, and the use of new technologies shaped housing production.

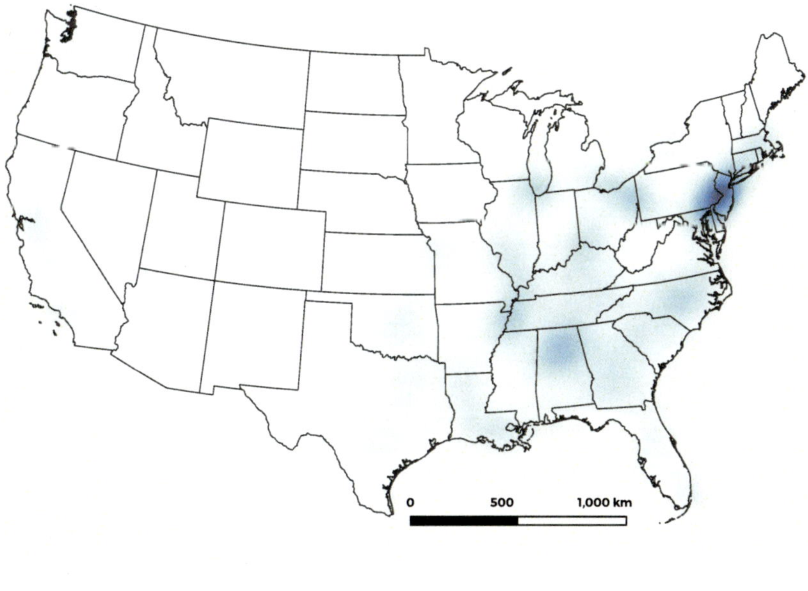


**FIGURE 2.9** Heatmap of people living in public housing

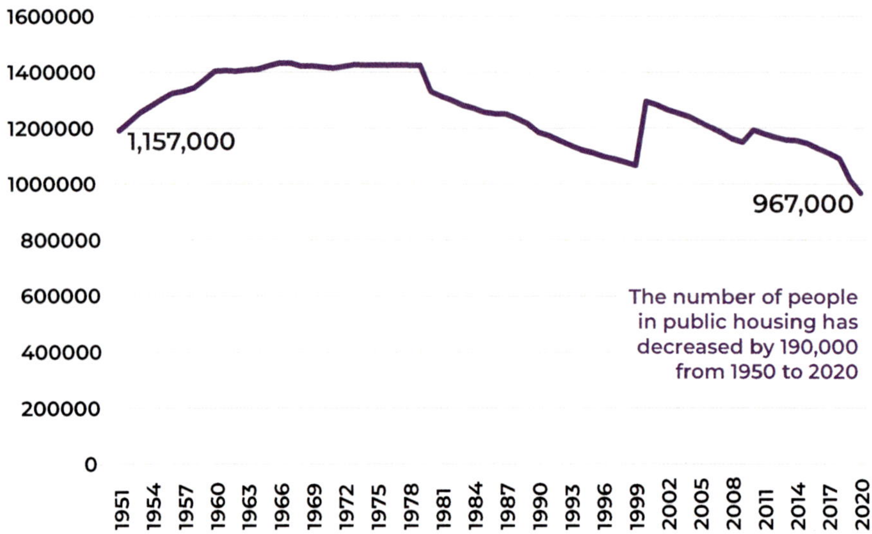


**FIGURE 2.10** Number of people in public housing over time

### 2.2.1 Fire

In October 1871, the Great Chicago fire ripped through downtown Chicago killing 300 people and left one-third of the population homeless. The fire burned for 24 hours and destroyed 17,500 buildings (National Geographic, 2022). At this time, the construction of most buildings in Chicago and other cities such as New York and Boston used wood-frame construction. After the fire, laws were put in place to construct buildings with fireproof materials, but many could not afford the materials and simply did not follow the new laws. In 1874 another fire destroyed 800 buildings in Chicago and finally new buildings began to follow the regulations for construction with fireproof materials. The buildings that were constructed of cast-iron were just as vulnerable as the wood constructed buildings. After this latter fire and the discovery that terracotta could protect cast iron construction in fire, the regulations for fire protection in construction began to be followed. This pushed out residents from downtown Chicago that could not afford to build in these new materials and methods, changing settlement patterns for those with lower economic standing.

Similarly, in New York City tenement housing, housing with three or more dwelling units, was built to house immigrants coming to the United States. Housing advocates became very concerned about the conditions of these types of homes as early as the 1860s. Conditions inside each dwelling unit were such that only one room had direct access to light and air while all other rooms were windowless. The response to these conditions was the Tenement House Act of 1879. Jacob Riis photographed many of the conditions (Riis et al., 1890, see Figure 2.11). The act required windows in all rooms. Adherence to the requirement led to what is known as the dumbbell style tenement building (White and Willensky, 2000). As well as light and air, fire escapes and fireproof balconies and stairwells were required to prevent human loss during a fire. These changes did not address all the problems in tenement housing design. New York State Tenement House Act of 1901 was enacted to further improve housing conditions in tenements.

The 1901 Tenement House act incorporated requirements for light in rear yards as well as minimum separations for courts and requirements of bathroom facilities.

This law also retroactively imposed restrictions on old law tenements for bathroom facilities and increased lighting. The 1901 act also required new and old tenement buildings to install fire escapes. This regulation was strictly enforced, and the visual landscape of NYC began to change. The tenement houses of this period take the shape of letters, typically I, H and C, formed by the required courts between the buildings. This act sparked a spurt of development right before the law was passed because developers rushed to build before the new law went into place in order to skirt the new regulations. It also prompted developers to increase the number of building floors from 4–5 stories to 6 or 7 stories (without elevators).

Fire was always a major challenge. The Sanborn Map company made detailed and large-scale maps of major US cities that provided a great deal of information to assist insurance providers in assessing risk. These maps were created beginning in the late 19$^{th}$

**FIGURE 2.11** Jacob Riis, 'Bandits Roost'

century. Although they were created to assess the risk of fire for insurance purposes, over time their significance extended beyond this use. They became a resource for urban planning and development because they provided, as seen in Figure 2.12 (Sanborn Fire Insurance Map, 1898), comprehensive information about buildings, structures, streets, and infrastructure. The maps depicted the layout of cities and towns in great detail. Data included building materials, property widths, and the location of fire hydrants and were updated on a regular basis, generally every 2–5 years. These maps were meant to provide accurate and up-to-date information; therefore, urban planners were able to use them to gain insight into the past and analyze urban growth patterns.

If the Sanborn maps were accidentally deployed to serve planning purposes, the Public Land Survey System (PLS) was established in the United States with the explicit purpose of managing land with an intent to promote orderly growth. The PLS was used to

FIGURE 2.12 Example of a Sanborn fire map

survey and divide land in the western territories (more detail about the PLS can be found in Section 2.3). Using cadastral maps, detailed representations with land ownership, land use, and property boundary data were created. The maps included surrounding features such as roads, water bodies, and neighboring properties. Cadastral maps illustrate the spatial arrangement of land ownership, which had a significant impact on urban planning, providing spatial information that helps guide and inform planning decisions. The maps offer urban planners an understanding of existing land use patterns, identifying available land for development, and assessing the potential for urban expansion. By analyzing cadastral maps, planners can determine the suitability of different areas for specific land uses, such as residential, commercial, or industrial zones, see Figure 2.13. These maps also aid in identifying infrastructure needs, including road networks, utilities, and public amenities, by highlighting the spatial relationships between parcels and infrastructure.

### 2.2.2 Elevators

As new buildings rose in the late 1800s in cities like Chicago and New York, their heights were limited to how many stories a person could reasonably climb, typically around six-stories. In 1857, the Otis Elevator Company began manufacturing passenger elevators for tall buildings in New York City. These elevators first found a place in commercial buildings. Commercial buildings began to rise taller and taller with the elevator, and eventually became an issue because of the shadows they created below/around them. In 1916, the NYC Zoning Resolution addressed these issues of bulk for growing skyscrapers to mandate for light to penetrate the streets below.

At the time of the 1916 zoning resolution, wealthy New Yorkers were still living in townhouses consisting of only a few stories and adjusting to the idea of apartment

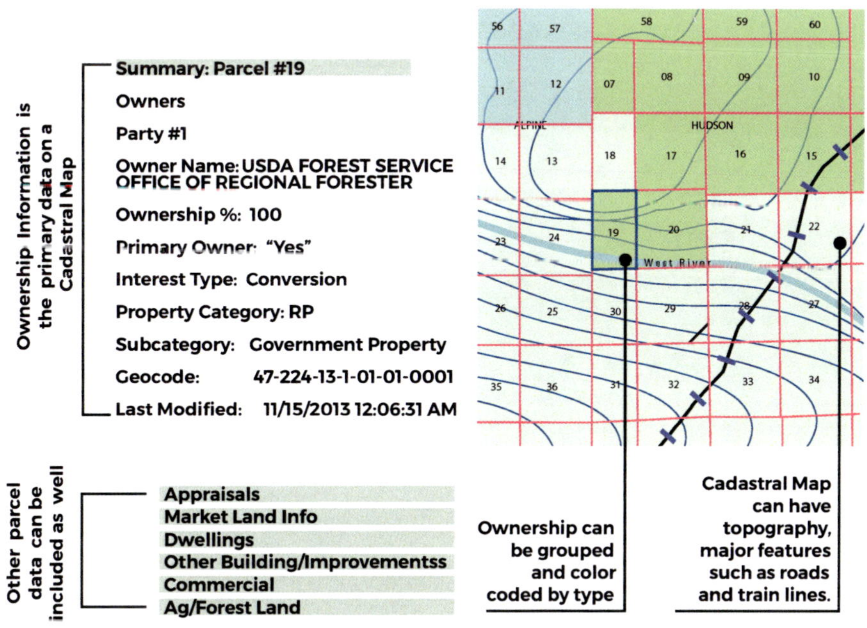


FIGURE 2.13 Example of a cadastral map

living, but by the 1920s the idea of living on a higher floor began to emerge as a status symbol and many high-end apartment buildings were built from then onward (see Figure 2.14 (data source, MapPLUTO, n.d.) and Figure 2.15 (data source, NYC Open Data Portal, 2022).

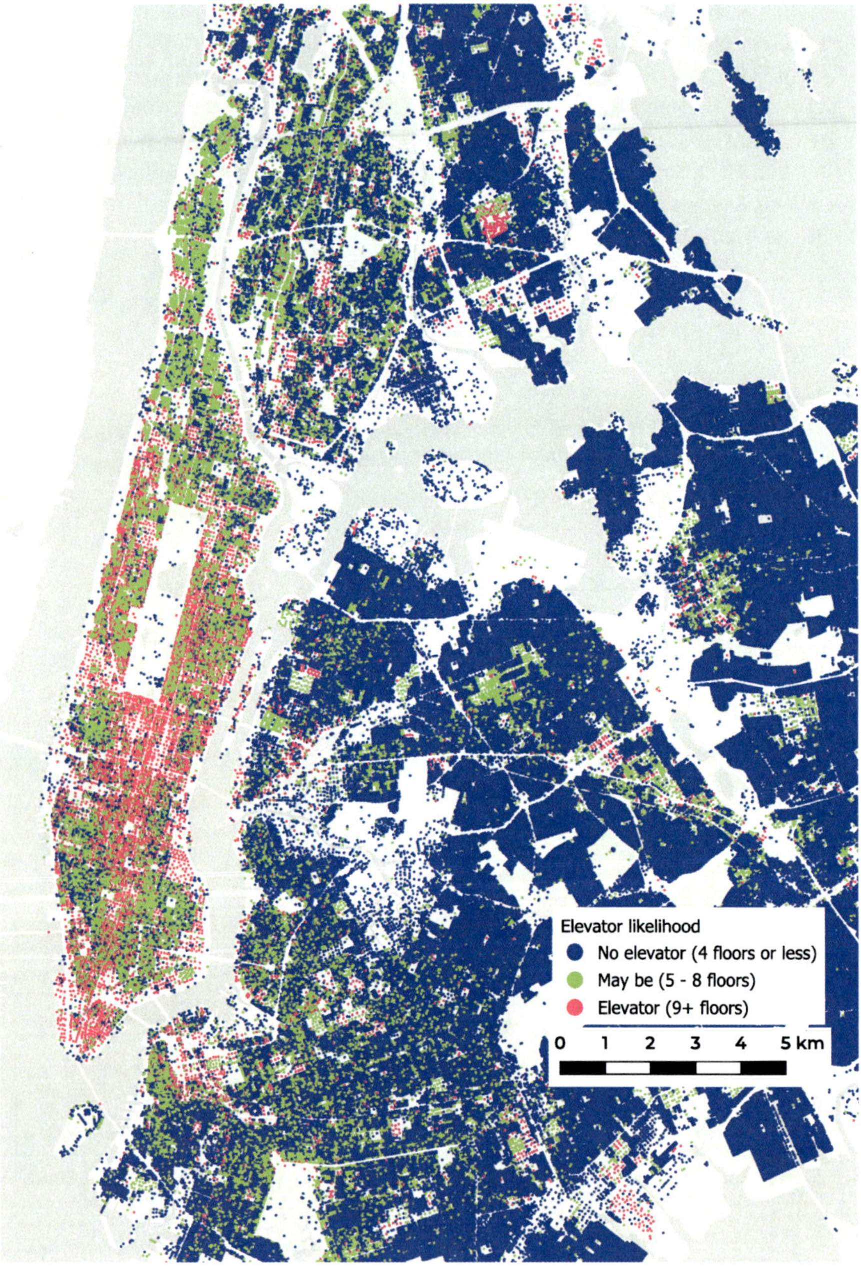


FIGURE 2.14 Vertical exploration: mapping Manhattan's elevator distribution

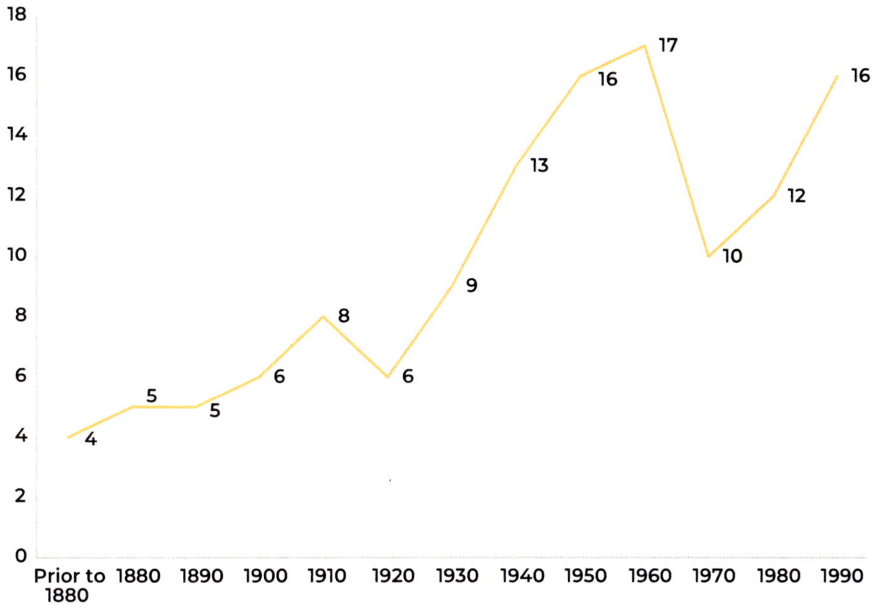


**FIGURE 2.15** Decadal analysis of Manhattan's average building height

It wasn't until the late 1940s during urban renewal that middle- and low-income housing began to take advantage of elevator buildings. The forms of the low-income housing projects that were built were also a product of the zoning resolution that allowed buildings to build vertically as long as open space on a lot was maintained. These "towers in the park" were developed not only in NYC, but in almost all US cities during this time period.

### 2.2.3 Air Conditioning

Air conditioning also made the rise of the skyscraper possible. Once a building reaches greater heights, operable windows are not reasonable because of high winds at those elevations. With the implementation of air conditioning in tall commercial buildings, windows were no longer needed for air, and could always remain fixed in place. Air conditioning and elevators allowed for skyscrapers to rise as tall as the structural system would allow. The 36-story Philadelphia Saving Fund Society (PSFS) building, built in 1932 in Philadelphia, PA was the first international style skyscraper in the United States that used air conditioning for ventilation and comfort for the commercial tenants in lieu of operable windows (see Figure 2.16 (en.wikipedia, 2023)).

Air conditioning was introduced to Americans in commercial settings in the early 1900s. The company Carrier, a maker of fans at the time, developed air conditioning to lower humidity in printing factories during hot summer months in New York. There was great success, and the technology quickly became a standard in factories of that time. In the South, textile mills and tobacco processing plants also employed the use of air conditioning, not so much for the workers, who did benefit from its use, but for the manufacturing of the product. It wasn't until the 1950s that air conditioning

**FIGURE 2.16** Philadelphia Saving Fund Society (PSFS) building on Market Street

became affordable enough to be marketed to the general population. Residential air conditioning had a great impact in hot areas of the South and Southwest United States (see the discussion about the growth of Phoenix in Section 1.4.3).

In addition to the growth of industry, people were able to live in these areas because of the increased comfort that came with air conditioning. As of 2015, all new housing in the South has central air conditioning, emphasizing the importance of conditioned air for comfort in this region. Air conditioning is now a standard in most new homes, but the flipside of that indoor comfort means that these regions now spend as much energy on cooling as more northern climes spend on heating during the winter, thereby contributing to global warming (see Figure 2.17 (USGCRP, n.d.)). New technologies are being developed to make air conditioning systems more

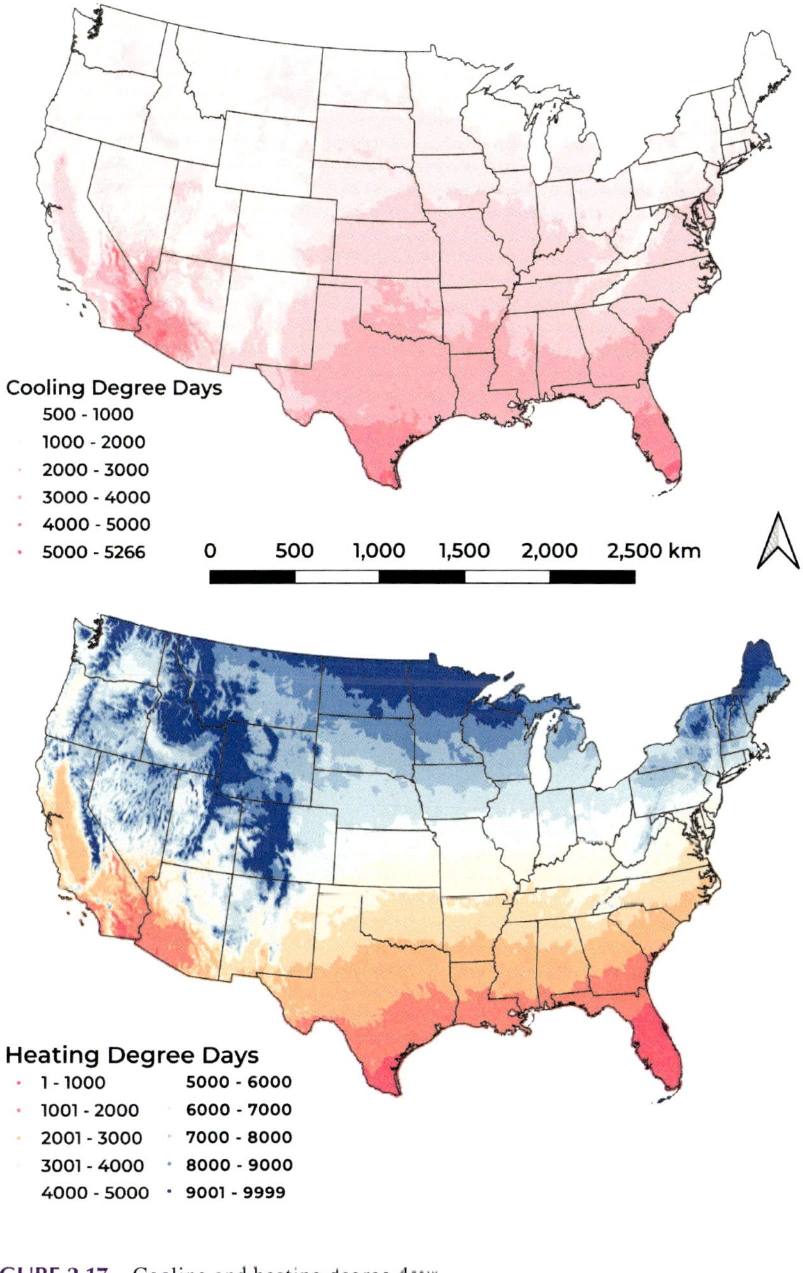


**FIGURE 2.17** Cooling and heating degree days

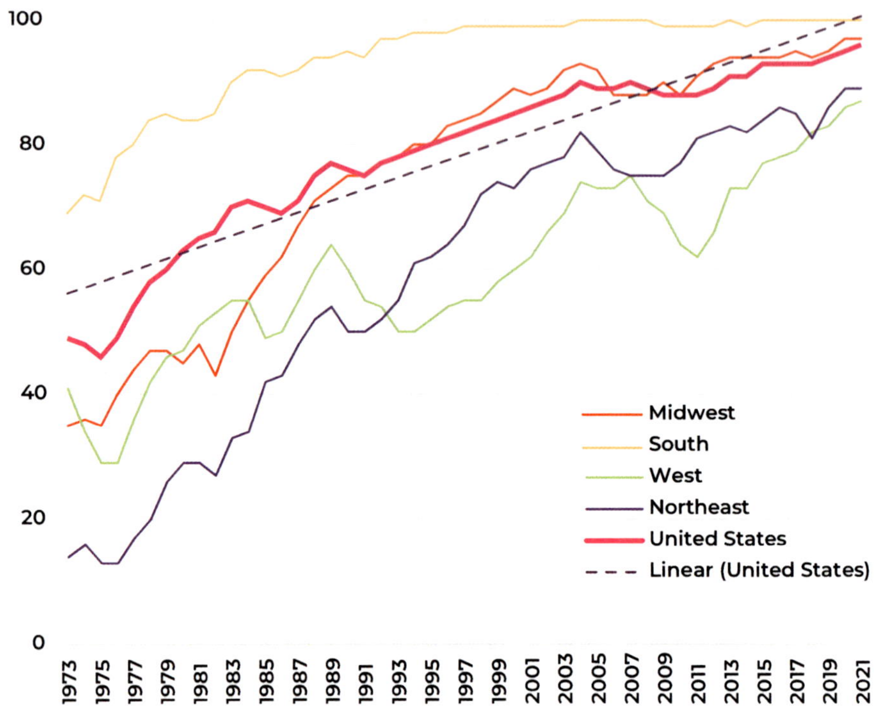


**FIGURE 2.18** Cooling the nation: Comparative regional analysis of air conditioning adoption in new single-family homes from 1973 to 2021

efficient overall. Over time the use of air conditioning has increased and now there is a strong push for architects and designers to design buildings for thermal comfort using passive methods to reduce cooling or heating loads on a building (see Figure 2.18 (data source, US Census Construction, 2023)).

### 2.2.4 Transportation

The early stages of the American industrial revolution created innovations in transportation technologies, specifically a shift away from water-based transportation from the East Coast through the Great Lakes towards the development of rail-based transportation. Regionally, the expansion of the railroads opened up the western United States creating opportunities for settlements to emerge along these newly established train routes.

Subsequently, the development of the automobile to support private transportation changed the form of our cities. The "walking city" of the late 1800s gradually gave way to the "streetcar city" of the early 1900s (Schiller and Kenworthy, 2017). By the 1920s, American cities began to expand to accommodate the private automobile, which could move people of affluent means away from the congested and unsanitary city to the bucolic countryside. In New York, master planner Robert Moses created scenic "parkways" to create a pleasurable experience for those who traversed in automobiles, creating opportunities for the journey to be as pleasurable as the destination (Caro, 1975).

This system of parkways played a significant role in shaping The Bronx, initially because most of the parkways ran in a north-south direction from the wealthy suburbs in the north to Manhattan in the south (see Figure 2.19, Nelson, 2023). In the undulating geographic terrain of The Bronx, the roads were like rivers that run along

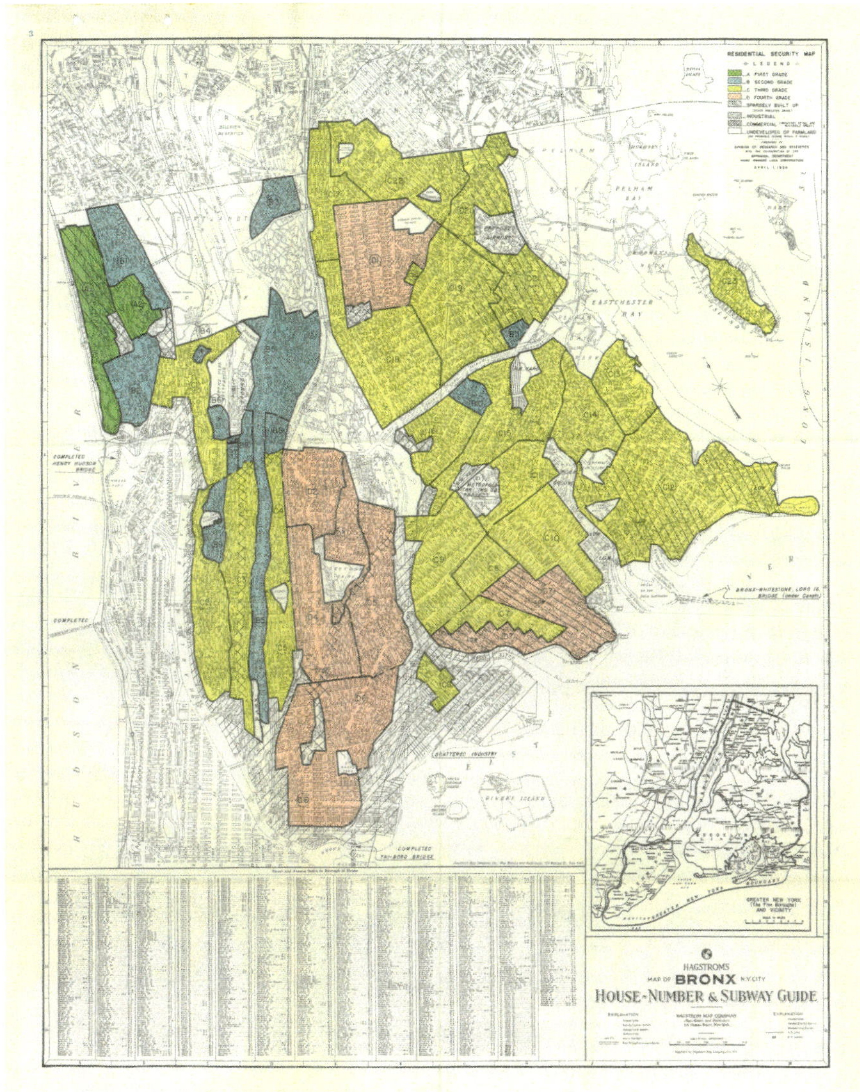


**FIGURE 2.19** Bronx HOLC map

the valleys. Likewise, the commuter rail lines were designed to move commuters from Westchester and south-eastern Connecticut directly into Manhattan, then considered the nerve center of the metropolis. These rail connections largely bypassed The Bronx, just as the parkways did.

In addition to the rail and road transportation developments, another 20$^{th}$ century federal government supported intervention was also influential in creating and establishing patterns of neighborhood settlement and displacement. A New Deal program called the Home Owner's Loan Corporation (HOLC) attempted to assess and ameliorate the problems faced by homeowners in the wave of the Great Depression. Real estate professionals created "residential security" maps to classify neighborhoods where examiners systematically graded neighborhoods based on criteria related to the age and condition of housing, transportation access, closeness to amenities such as parks or disamenities like polluting industries, the economic class and employment status of residents, and their ethnic and racial composition. Neighborhoods were color-coded on maps: green for the "best", blue for "still desirable", yellow for "definitely declining", and red for "hazardous". Figure 2.19 (Nelson, 2023) is an example of a HOLC map from that era.

In the post-war era, freeway placements and expansions in urban areas typically occurred where land prices were depressed, which frequently corresponded with the residential neighborhoods of low-income and minority households.[2] Such neighborhoods generally had low levels of political power resulting from institutional discrimination over time. In some respects, freeway locations in cities are the philosophical progeny of "Negro removal" or "urban renewal" programs that were thought to cure "urban blight" by tearing down minorities' homes (Powell and Graham, 2002). More than 200,000 people have lost their homes nationwide to federal road projects over the last three decades, according to a Los Angeles Times analysis of federal transportation data (Dillon and Poston, 2021).

Figure 2.20 (Google Earth, 2022) shows an aerial view of Link Road in the Independence Heights neighborhood of Houston where a mural was painted to highlight the uniqueness of Independence Heights, one of the oldest Black communities in Texas (Hennes, 2020).

## 2.3 LAND USE PLANNING

Land planning in the United States began as the country became settled and several governmental entities were actively involved in land surveying and classification. The first major survey of public lands was initiated in 1785, see Figure 2.21 (Bechler et al., 1856). The Public Land Survey system, also known as the rectangular survey system, was first proposed to commodify "public lands" to build a new nation. American land surveyors divided the land into sections of one square mile containing 640 acres. Townships consisted of 36 sections on a rectangular grid (Bureau of Land Management, 1991).

While the Bureau of Land Management is responsible for the management of public lands, the surveying procedures adopted over two centuries ago continue to shape how land is measured and mapped even today.

**FIGURE 2.20** Houston Black community aerial perspective

The term "land use" sometimes written as "land use" can be interpreted as a simple descriptor that explains how the land around us is being used. However, many professions including surveying, architecture, urban planning, and engineering define land use through the lens of commodifying and classifying land in order to shape development and growth. The main considerations in discussing land use patterns are the concept of land value, and inherent in the assessment of land use is the belief that land should be used to its highest and best potential.

The American Planning Association describes the national land use classification schema that is used in the United States. Land Based Classification Standards consider different variables that describe a land parcel – including observable activity (e.g., farming or manufacturing), economic function (e.g., agricultural, commercial, or industrial), structure (e.g., single-family home or office building), site (physical characteristics that can help to assess whether the land has development potential or not), and ownership (identifying who has the rights to develop the land). The LBCS also includes a detailed color-based classification that is deployed across all land use maps although local variations may exist. In general, residential activities are coded yellow, commercial activities are coded red, institutional activities are coded blue, agricultural activities are coded green, and unclassified land is coded white.

Contemporary land use maps are created by taking data that describes the use of a parcel or piece of land. The use is classified into categories. The scale and type of land being examined determine the categories that will be shown on a land use map. A regional land use map may show built-up urban land use and agricultural or forest land. Land use maps that are at a city level can show open space or parks

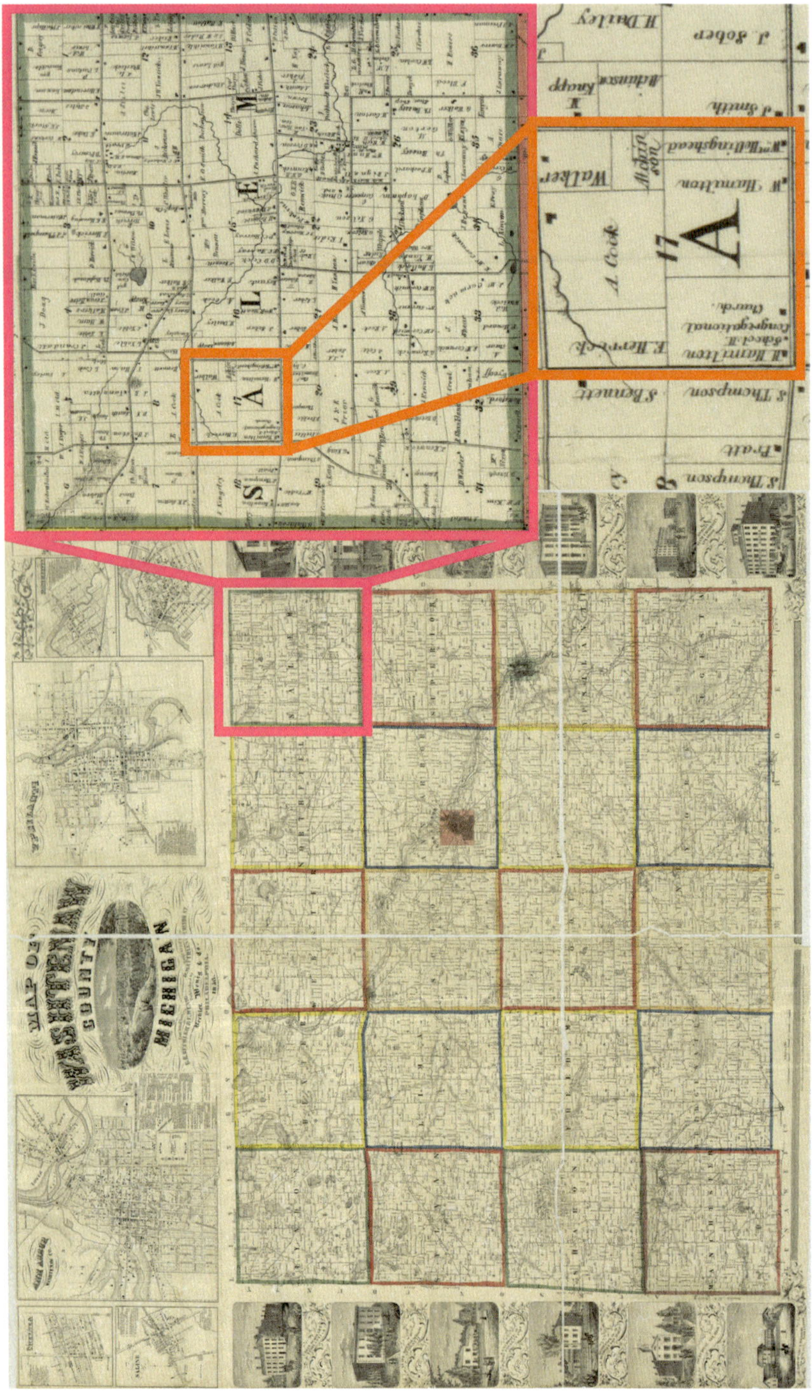


**FIGURE 2.21** Public land survey system: historical map of 1856 Washtenaw County, Michigan

**FIGURE 2.22** NYC ZoLa land use map

and recreation areas, residential areas, commercial areas, and manufacturing areas. Maps that are at a neighborhood or block level will give further details into the land use, for example, showing detailed information about residential land use, such as multi-family use vs. single-family housing (see Figure 2.22 (NYC Planning, 2023)).

Such land use maps help us to observe patterns like higher concentrations of multi-family housing near transit, or manufacturing near a waterway. This type of analysis can help housing professionals understand the spatial patterns of housing and the past and potential impacts of the policies created on spatial makeup of a place.

## 2.3.1 Land Use and Zoning

Land use maps describe the characteristics of the land in its present state, whereas a zoning map codifies present land uses, considering societal needs and aspirations. While an expanded discussion of the history of zoning in the United States and around the world is beyond the scope of this book, zoning, in its simplest form, is the creation of single-purpose districts or "zones" where one particular type of land use/activity can occur (Hirt, 2014). Traditional zoning formalizes present and future land use, regardless of ownership. Zoning considers public health (access to light and air), safety (avoiding overcrowding), pollution (the separation of heavy industrial activities away from residential living areas), as well the need to provision space for desirable uses such as parks and playgrounds. The concept of a rigid separation of uses is a vestige of the City Beautiful movement (Hall, 2014).

In the United States, planning is highly localized (Hoch et al., 2000). Zoning supports planning and is a powerful instrument that transforms a local government's political visions into reality. For example, a local government that wants to increase its property tax revenues may zone a higher proportion of available land for single-family housing than for multi-family residential housing (rental apartments or condominiums).

The earliest reforms of housing focused on the design of individual multi-family dwelling units, the tenements discussed in Section 2.2.1. The 1901 Tenement Law required an interior courtyard for ventilation and garbage removal, rather than relying on interior air shafts that could not be cleaned. Additional requirements and improvements focused on indoor plumbing and removing waste and connecting tenements to a sanitary sewer system. By the early 1900s, the City Beautiful movement was growing in western societies including America, and well-meaning elites advocated for a benevolent way to manage the housing needs of the masses.

Zoning is a set of regulations and restrictions that municipalities impose onto private properties. These laws began with Los Angeles in 1904 and New York City in 1916, in a continuation of the efforts to improve sanitary conditions described in Chapter 1, Section 1.3.1. At this time, it was a new idea that private owners could have restrictions on what they could build on their land, not only in size but in use. In 1926, a case against zoning was brought to the Supreme Court in the Village of Euclid v. Ambler Realty Company (see Figure 2.23 (Kull, 2023)). This case cemented a local government's right to impose zoning restrictions upon land based on the notion that there was a right to maintaining the character of a neighborhood. After this ruling, there was an increase of zoning regulations implemented in the United States, see Figure 2.24 (data source, APA, 2023), which shows the year of implementation of zoning regulations for the largest city in each state, most of which fall between 1920 and 1930.

New York City passed its first zoning regulation in 1916. This document was among the first of its kind and regulated the height, use, and lot coverage of buildings. They addressed issues such as undesirable shading of neighboring streets as well as the desire from wealthy residents to keep the encroaching manufacturing uses away from Ladies Mile, which at that time was a posh shopping district. The second zoning resolution in NYC was passed in 1961 to include the separation of all buildings into three use zones, commercial, manufacturing, and residential.

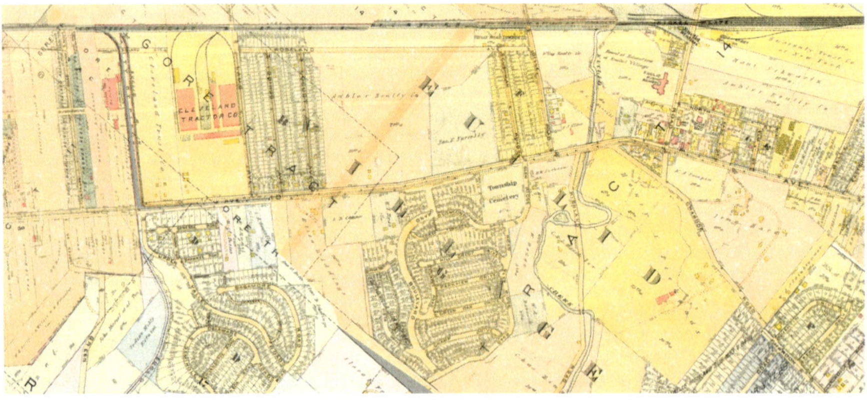

FIGURE 2.23 Euclid vs. Ambler – eminent domain

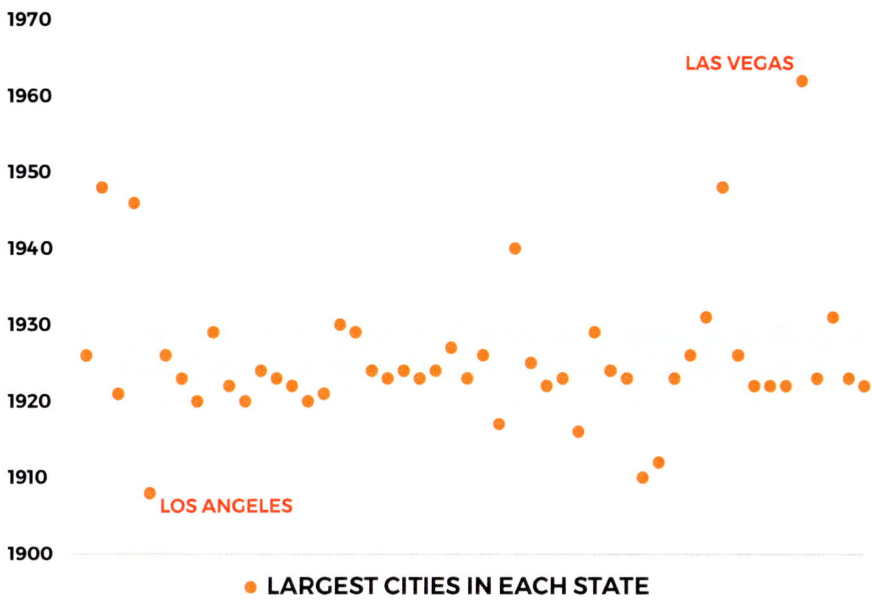


**FIGURE 2.24** Urban regulatory evolution: chronology of zoning code enactment in major US cities

### 2.3.2 Redlining

Redlining can be defined as a discriminatory practice that consists of the systematic denial of services such as mortgages, insurance loans, and other financial services to residents of certain areas, based on their race or ethnicity. The term redlining finds its origins in the HOLC program previously described in Section 2.2.4. These maps were color-coded, each color corresponding to the loan worthiness of the neighborhoods in the United States and the color red was attributed to the neighborhoods that were deemed not worthy of inclusion in the homeownership programs. Most of the neighborhoods marked in red were predominantly inhabited by Black residents. The consequences of this were that Black residents were denied government-insured loans.

The University of Richmond's Mapping Inequality project digitized scans of an example of such redlining maps developed by the Home Owners' Loan Corporation (HOLC), which it is important to note did not engage in redlining (Gomby, 2022). Nowadays, about 11 million Americans live in those formerly red-zoned areas. This population is now majority-minority but not majority-Black, nor do Black residents form a plurality in these areas overall. The Black population share is approximately 28%, ranking third among the racial groups who live in formerly redlined areas, behind White and Hispanic residents. The approximately 3 million Black residents in redlined areas account for just 8% of all Black Americans.

As discussed in Section 1.3, there are great regional differences in the effect of redlining today, which would be better characterized as the persistence of sustained racism. Only 7% of the population in formerly redlined areas in Denver are Black, while some 85% of the 80,000 redlined residents in Birmingham, AL, are. As many inner cities are gentrifying, Black-majority suburbs are on the rise (Saunders, 2019), which were underrepresented in HOLC maps due to their focus on urban centers.

In the 1990s, another form of redlining became apparent as homeowners who lived and owned properties in certain "redlined" census tracts that were dominated by Blacks or people of color did not receive the same homeowners' insurance products as those who lived in predominantly white census tracts. Although Milwaukee in the 1990s was a spatially segregated city, the segregation can be masked if the data is analyzed at the level of zip codes (a larger area) that can mask intentionally discriminatory practices. Figure 2.25 shows a finer resolution that begins to show the spatial correlation between insurance policies and African American communities. In Milwaukee, the work of nonprofit groups and legal activism fostered a settlement with a large insurance company who systematically discriminated against African Americans (Ramasubramanian, 1995).

Zoning has been criticized by the political right for creating a vast array of rules and regulations that contravene private property rights and by the political left for serving powerful interests by zoning out "undesirable" (less profitable) uses (Angotti and Morse, 2023). More recently, zoning has come under rigorous scrutiny and withering criticism as analysis of historical zoning maps has revealed a more deliberate attempt to create racial segregation than previously thought (Rothstein, 2018).

Larger cities like New York have "rezoned" land, to create new opportunities as traditional land uses have ceased to exist. Formerly industrial areas have been rezoned to allow a range of uses including residential and commercial uses. In New York City, prized industrial waterfront land has been made available to developers who have created market-rate residential living units or other kinds of luxury commercial ventures that cater to tourists and the elite.

New York City has been at the forefront of planning since the area was first settled in 1609 (Sanderson, 2009). As the city and region grew rapidly from the 1600s to the 1900s, the city's leaders encountered challenges related to managing the built environment. The naturally occurring geographies of the settlement patterns meant that while social classes lived next to each other, their lives and lived experiences were anything but similar. In addition, self-selection based on ethnicity, country of origin, and/or mother tongues prompted the emergence of distinct residential enclaves that have persisted over decades. Yet, Little Italy in NYC remains a distinct enclave in name only, a physical vestige and a landmark reminding us about the complexities of neighborhood change and assimilation.

## 2.4 SUBURBANIZATION AND URBAN SPRAWL

Suburbanization refers to the socio-spatial process whereby cities expand outwards beyond their original central areas via the formation of suburbs. Suburbs are peripheral areas lying beyond a city's boundaries, but which are interconnected to the city economically and socially, for example, via commuting. Suburbanization typically involves building new homes for either sale or rent, combined with residential mobility whereby people leave the city in order to live in non-urban settings.

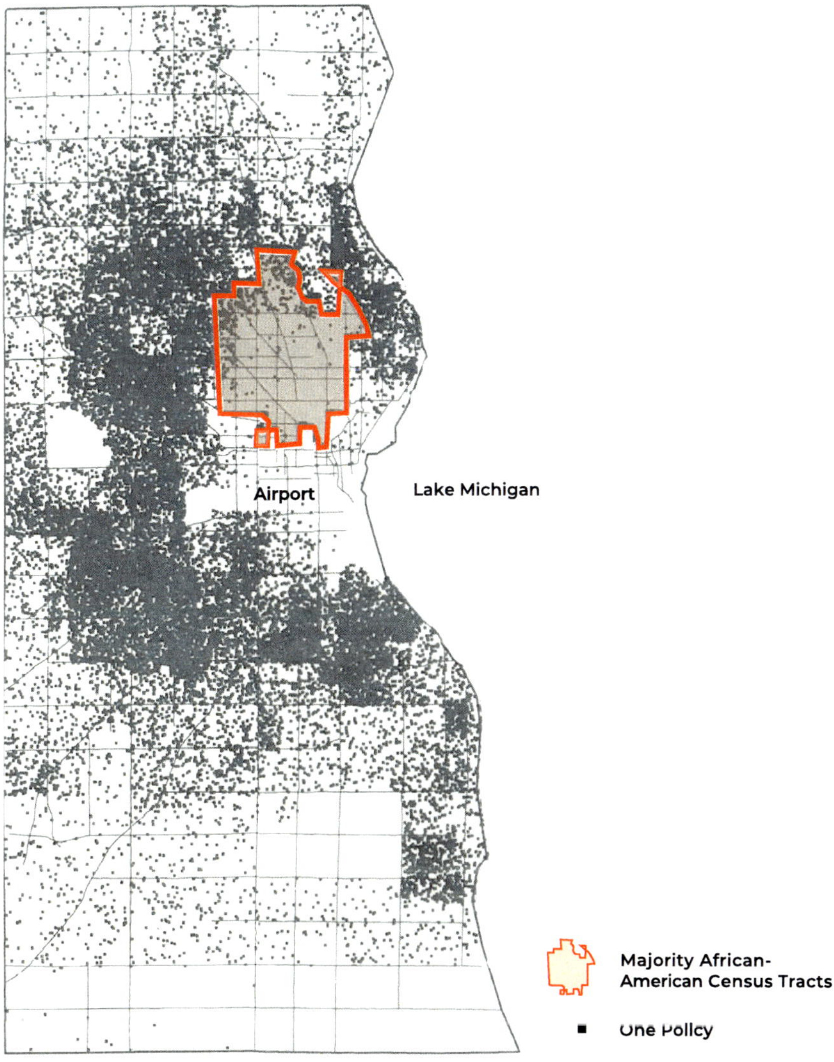


FIGURE 2.25 Unveiling the diversity of Milwaukee's urban landscape: an exploration of one-policy areas and African-American majority census tracts

### 2.4.1 First and Second Ring of Suburbs

Suburbs used to be defined by their relationship to the urban core area that they surround. This started to change in the late 1990s, when increasing areas in the US South and West that have no more than villages or small agricultural towns at their center began morphing into suburban corridors. Morphologically and socially, these areas act as suburbs, even if they do not have urban centers (usually defined as areas with at least 50,000 residents and more than 1,500 residents per km$^2$). The extent of suburbs has been steadily increasing throughout the 20th century and in some parts of the country into the 2020s. Multiple factors have been playing changing roles in this development. As cities grew, greater numbers of "inner city" residents sought to escape the core areas, whose infrastructure was not designed to cope with the densities created by the explosion of urban populations. The first ring of suburbs was facilitated by the advent of streetcars and suburban railway systems in the first half of the 20th century. A second ring of suburbs was accommodated by the general availability of the automobile after the Second World War. Both developments were accompanied by a depopulation of rural areas, while the second ring of suburbs was also fed by (mostly White) residents fleeing deteriorating inner cities.

Starting in the (late) 1990s, as those inner cities began to re-gentrify and the infrastructure in the first ring started to deteriorate, the two populations began to replace each other. These phases played out at different times in different parts of the country until the Covid pandemic of 2020 introduced completely new settlement trends. First ring suburbs are structurally different from second ring suburbs. The former is older and denser and requires an urban core (Puentes and Warren, 2006); the latter is post Second World War and often much younger than that, has typically no rail infrastructure, and is hence car-dependent, which in turn leads to lower population densities and a lack of focus on urban functionality. See Figure 2.26 (US Census, n.d.; NHGIS, n.d.) which maps the suburbanization of Dallas over time.

### 2.4.2 Edge Cities

As suburban lifestyles became the norm in the United States (European and Asian cities have a different trajectory because of cultural and space constraints), suburbs became less and less dependent on an actual metropolitan center and developed as second ring suburbs both spontaneously as well as in the form of planned edge cities. Edge cities in the narrow sense of the term's inventor Garreau (1991) formed around office parks or shopping malls, which replaced the core that used to be the necessary ingredient for first ring suburbs. Individual, automobile-based transport, and an often politically motivated disdain for cities, resulting in preferential treatment of the usually White population in second ring suburbs, together with the availability of large and relatively cheap tracts of land quite literally paved the road for large swaths of formerly agricultural land to be transformed into low-density residential areas with no discernable boundaries (Firestone, 2001). Driving through those second ring suburbs in the Sun Belt or California, one is hard-pressed to see where one community ends and the next one begins. These areas are the epitome of sprawl. Figure 2.27 (US Census, n.d.; NHGIS, n.d.) highlights the relationships of densities changes from the 1980s.

### 2.4.3 Urban Sprawl

Urban sprawl is characterized by the lack of coordination among the communities within which it occurs. Associated with this is a lack of concern for the consequences leading to unsustainable living conditions as people age without having access to services for the elderly, and energy costs skyrocket. Another argument is

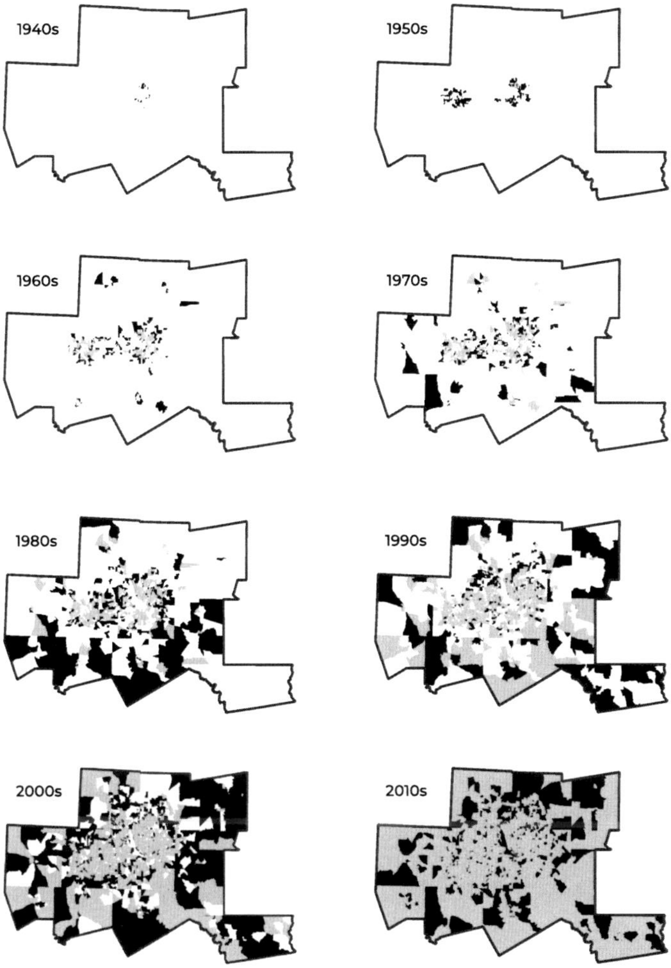


**FIGURE 2.26** Phenomenon of suburbanization in Dallas, Texas

that the affected communities, similar to what we discussed for rural disadvantages in Section 1.3.5 of Chapter 1, cannot afford to provide and maintain the necessary infrastructure, or if they do, engage in a social sorting as only wealthy residents could afford the higher costs associated with lower densities. Research into the effects of urban sprawl has drawn the attention of public health scholars as the number of traffic accidents, obesity, and diabetes rates has been shown to have a positive and significant relationship with urban sprawl (e.g., Frumkin et al., 2004).

The phenomenon has received widespread attention within the planning community (e.g., Oliver, 2002; Squires, 2002). Several measures to define and determine the intensity of the phenomenon have been developed and debated (e.g., Ewing and

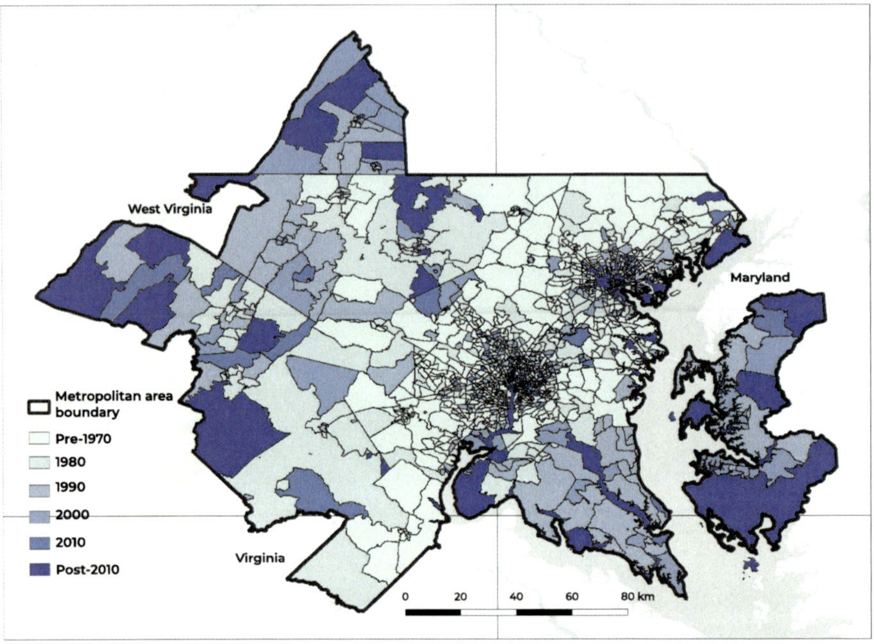


FIGURE 2.27 Metro Washington, DC growth from 1970 to 2020

Hamidi, 2014). In one way or another, all measures center around residential and job density, the distance between sites of human activity, and spatio-temporal measures of accessibility. We will discuss the sources, combination, and compilation of such measures in Chapters 4–6; let it suffice to state here that GIS is essential to the development/validation of these measures as it is the specific spatial configuration of factors that determines the effects of sprawl.

## 2.5 GENTRIFICATION

In our discussion of suburbanization, we mentioned the reverse movements of people living in inner cities and those who live in the first ring of suburbs. As cities turned economically around and started to become more attractive again, the demand for housing started to rise, placing financial pressure on those who had remained in city centers. The replacement of local populations by deeper pocketed ones is known as gentrification. These kinds of replacement processes have been occurring throughout the history of urban development and may as such be considered "natural". Urban planners are in the inevitable position that the very policies aimed at revitalization then also lead to the displacement of people who cannot afford the rise in rents that follow the improvements – at least in a market-oriented society.

Gentrification is a highly politicized topic and its effects have often been exaggerated. Most neighborhoods in the limelight of political discussions have not actually experienced displacement (Freeman, 2005). Instead, the "gentrifiers" move into additional units, increasing the population density rather than replacing existing residents. This is not to say that displacement does not occur – but it happens at a much lower rate, and in many places not at all, than the process is maligned for. Part of the misconception is the general rise of unaffordable housing (see Chapter 1), an experience that gentrifying neighborhoods share with everybody else. A comparison of affordability rates in 1970 with those in 2020 shows only five rural counties (out of over 3,000), where housing has become cheaper relative to the median income in the respective county (three of those counties have fewer than 5,000 residents).

We alluded to the fact that residential change is a given; even if functionally, or social status-wise, a neighborhood remains similar to itself, the people living in those neighborhoods tend to change. New York's Little Italy and Milwaukee's Germantown are monikers for bygone eras whose residents now show little resemblance with the neighborhood's namesakes. The role of the planner is then to prevent deterioration (which usually requires collaboration with other city departments), work continuously on improving conditions, and smoothen transitions as the inevitable change is taking place. Neighborhoods find themselves in the crosshairs of multiple processes inside (aging populations) and outside (suburbanization and its reversal), but sometimes, individual events or actors may play an outsized role. Urban universities have acted as such actors of change, where growing student bodies and massive technology investments have been forces of gentrification. In the age of knowledge work, college towns and their equivalent in an urban context have been engines of growth, which if not handled carefully, can indeed result in the displacement of small local businesses and less adaptable residents. As such investments are mostly in non-residential buildings, rental costs (or home ownership) rise disproportionally leading to an additional squeeze in a housing market that is characterized by the phasing out of rent restrictions.

Zoning has been hailed (Schuetz, 2019) and vilified (Angotti and Morse, 2023) as the cause for the limited availability of housing and hence the replacement of long-term residents by those who can afford higher rents. Where zoning limits density (see also Section 3.2.6), it certainly contributes to a housing shortage – but it is hard to argue that this is a cause for gentrification. Where zoning changes allow for residential units in what was formerly non-residential, it will certainly increase housing supply – but not necessarily in the affordable range. By definition, no gentrification could take place here as these areas had no residents before. Where zoning changes from purely residential to mixed use, especially with support of public transit (such as light rail stops), it is likely to increase the attractiveness of a neighborhood, which in turn is likely to raise property prices – this is one of the conundrums that urban planners have to live with. Case studies showing that such investments lead to actual displacement, however, are rare.

Figure 2.28 (US Census, n.d.) shows a correlation between the arrival of new residents in a census tract and reduced affordability for renters. In other words, when neighborhoods are perceived as desirable, either because of their trendiness, accessibility, or affordability, newer affluent renters move in, causing spikes in the rental market. Long-term tenants in these neighborhoods are displaced in favor of those who can pay more.

FIGURE 2.28 Dynamic Brooklyn (NY): Mapping rent increases and neighborhood changes revealing the relationship between rent and resident mobility with density reflecting recent census tract inflows

Up-zoning (increasing heights and easing restrictions such as parking minimums) has its advantages; in The Bronx, a planned growth strategy has encouraged infill development along major transit corridors (see Figure 2.29 (MapPLUTO, n.d.; MTA, n.d.). Although the bulk of the new development is in The South Bronx (closer to Manhattan), the map indicates that overall, The Bronx is seeing the benefit of planned rezoning in terms of the increase in new housing supply.

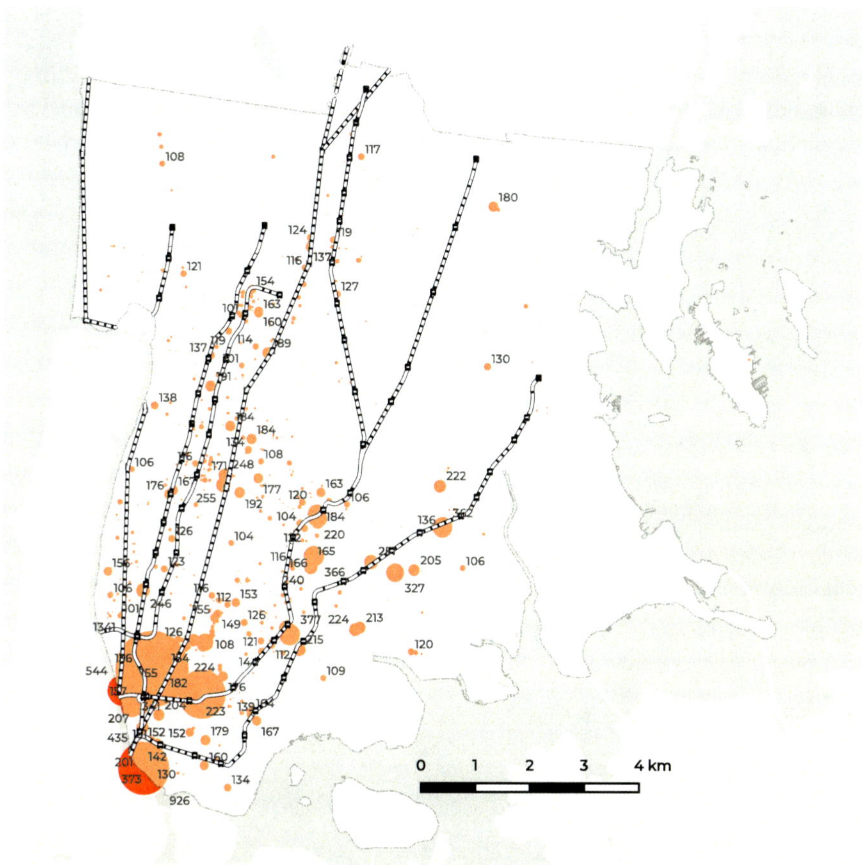


**FIGURE 2.29** The new Bronx

## 2.6 THE LIMITS OF ZONING

This brief historical run through focuses on understanding the demographic and technological shifts that shaped urbanization and suburbanization in late 19th and 20th century America. While not the main focus of this book, this chapter serves to remind GIS specialists about the complex social and political histories associated with zoning, not to mention its racist and exclusionary overtones that have disenfranchised and harmed African American communities and people of color. For GIS specialists, zoning is probably nothing more than a base layer of data that can be used to support complex analyses. While this is true, zoning is also an instrument that imposes a variety of land use controls that can empower or harm the lives of everyday people especially in contested spaces. On either side of the political spectrum, affordable housing activists and commercial housing developers will claim that restrictive zoning delays housing production, increases production costs, thereby reducing affordability, and creates exclusionary up-market residential enclaves. While it is easy to blame zoning for everything that is wrong with the housing situation, zoning is often the practical resolution of a value conflict – representing a compromise between preservation and development, between low and high density, and between having a single set of uses in a neighborhood vs. having a mixture of sometime incompatible uses in a neighborhood. This list can go on. Zoning cannot be relied upon as the only way to create and support a robust pro-housing agenda. In Chapter 3, we present a range of design and policy innovations that spur the development of just and sustainable housing options.

## NOTES

1. "Rent control" has a different connotation in Europe than in the United States. The ceiling is not hard and there are better established negotiation mechanisms, especially for larger multi-family complexes.
2. This was less prevalent in Texas and west of the Rocky Mountains as these states still had plenty of undeveloped land.

## REFERENCES

American Planning Association (APA), 1994. *Land Based Classification Standards*, LBCS. https://www.planning.org/lbcs/, last accessed 6 March 2023.

American Planning Association (APA), 2023. *Planning History Timeline*. https://www.planning.org/timeline/, 21 October 2023

Angotti, T, and Morse, S, 2023. *Zones Out! Race, Displacement, and City Planning in New York City*. New York: New Village Press.

Bechler, Gustavus R., Wenig, E. and Bechler & Wenig. Map of Washtenaw County, Michigan: from actual surveys. Philadelphia: Bechler, Wenig & Co, 1856. Map. Retrieved from the Library of Congress. www.loc.gov/item/2012593013/

Bloom, N, Umbach, F, and Vale, L, 2015. *Public Housing Myths: Perception, Reality, and Social Policy*. New York: Cornell University Press.

Caro, R, 1975. *The Power Broker: Robert Moses and the Fall of New York*. New York: Vintage.

Cavanaugh, E, 2020. *The Routledge Handbook of the History of Settler Colonialism*. Philadelphia, PA: Routledge.

City of New York, 2022. *NYC Open Data Portal*. https://data.cityofnewyork.us/Housing-Development/Building-Footprints/nqwf-w8eh, 19 March 2023.

Dillon, L, and Poston, B, 2021. "Freeways Force Out Residents in Communities of Color—Again". *Los Angeles Times*, 11 November. https://www.latimes.com/projects/us-freeway-highway-expansion-black-latino-communities/, last accessed 26 February 2023.

En.wikipedia, Absecon 59 at. "Wikimedia Commons." 1 1 2023. https://commons.wikimedia.org/wiki/File:PSFS-MarketSt.JPG, 19 October 2023.

Ewing, R, and Hamidi, S, 2014. *Measuring Urban Sprawl and Validating Sprawl Measures*. Salt Lake City, UT: Metropolitan Research Center. https://gis.cancer.gov/tools/urban-sprawl/sprawl-report-short.pdf, last accessed 2 June 2023.

Firestone, D, 2001. "90's Suburbs of West and South, Denser in One, Sprawling in Other". *NY Times*, 17 April. https://www.nytimes.com/2001/04/17/us/90-s-suburbs-of-west-and-south-denser-in-one-sprawling-in-other.html, last accessed 26 December 2022.

Freeman, L, 2005. "Displacement or Succession? Residential Mobility in Gentrifying Neighborhoods". *Urban Affairs Review*, 40(4): 463–491.

Frumkin, H, Frank, L, and Jackson, R, 2004. *Urban Sprawl and Public Health: Designing and Building for Healthy Communities*. Washington, DC: Island Press.

Garreau, J, 1991. *Edge City: Life on the New Frontier*. New York: Doubleday.

Gillham, O, 2002. *The Limitless City: A Primer on the Urban Sprawl Debate*. Washington, DC: Island Press.

Google Earth. "Google Earth." 14 January 2022. *Earth.google.com*. 29°49'00"N 95°23'30"W. 19 October 2023.

Gomby, G, 2022. "Reassessing HOLC Redlining Maps to Support Claims of Environmental Injustice". *Proceedings of the National Academy of Science*, 119(17). Doi:10.1073/pnas.2200211119.

Hall, P, 2014. *Cities of Tomorrow: An Intellectual History of Urban Planning and Design Since 1880*. West Sussex, UK: Wiley Blackwell.

Hennes, R, 2020. *Black Towns Matter' Mural Painted on Street in Houston's Independence Heights*. Chron. https://www.chron.com/houston/article/Black-Towns-Matter-mural-painted-on-street-in-15351923.php, last accessed 6 March 2023.

Hirt, S, 2014. *Zoned in the USA: The Origins and Implications of American Land-Use Regulation*. New York: Cornell University Press.

Hoch, C, Dalton, L, and So, F, 2000. *The Practice of Local Government Planning*, vol. 3. Washington, DC: International City County Management Association (ICMA) Press.

Holzner, M, and Huberman, M, 2022. "Red Vienna: A Social Housing Experiment." *The Journal of Interdisciplinary History*, 53(1): 49–88. Doi:10.1162/jinh_a_01798

Kull, A. *Property Visual Syllabus*. n.d. http://propertykull.weebly.com/village-of-euclid-v-ambler-realty-co.html, 21 October 2023

Lee, E, 2016. *The Making of Asian America: A History*. New York: Simon & Schuster.

MapPLUTO, n.d. *New York City Department of City Planning parcel database*. Online resource available at https://www.nyc.gov/site/planning/data-maps/open-data/dwn-pluto-mappluto

MTA, n.d. *New York Metropolitan Transit Authority*. Online resource available at https://open-data.cityofnewyork.us/data/

National Geographic, 2022. *The Chicago Fire of 1871 and the 'Great Rebuilding'*. https://education.nationalgeographic.org/resource/chicago-fire-1871-and-great-rebuilding/, last accessed 14 April 2023.

Office of Policy Development and Research. HUD User: Assisted Housing: National and Local. n.d. <https://www.huduser.gov/portal/datasets/assthsg.html>, last accessed 19 October 2023.

Perry, A, and Harshbarger, D, 2019. *America's Formerly Redlined Neighborhoods Have Changed, and so Must Solutions to Rectify Them*. Washington, DC: Brookings. https://www.brookings.edu/research/americas-formerly-redlines-areas-changed-so-must-solutions/, last accessed 29 May 2023

Pew Research Center, 2015. *Modern Immigration Wave Brings 59 Million to US, Driving Population Growth and Change Through 2065*. https://www.pewresearch.org/wp-content/uploads/sites/5/2015/09/2015-09-28_modern-immigration-wave_REPORT.pdf, accessed 1 June 2023.

Planning, NYC Department of City. *ZOLA New York City's Zoning and Land Use Map*. N.d.. zola.planning.nyc.gov, 13 July 2023.

Powell, J, and Graham, K, 2002. "Urban Fragmentation as a Barrier to Equal Opportunity". In Piche, D, Taylor, W, and Reed, R (Eds.), *Rights at Risk: Equality in an Age of Terrorism*. Report of the Citizens Commission on Civil Rights. Washington, DC: Citizens Commission on Civil Rights.

Ramasubramanian, L, 1995: "Building Communities: GIS and Participatory Decision Making". *Journal of Urban Technology*, 3(4): 67–79.

Riis, J, Bacher, O, Cox, K, Drake, W, Fitler, W, and Pérard, V, 1890. "Riis, Jacob A, photographer. *How the Other Half Lives: Studies Among the tenements of New York*. New York: Charles Scribner's Sons. https://lccn.loc.gov/04011775, last accessed 31 May 2023.

Robert K. Nelson, LaDale W., Richard M., Nathan C., et al., "Mapping Inequality". In. Robert K. Nelson and Edward L. Ayers (Eds.), *American Panorama*. https://dsl.richmond.edu/panorama/redlining/#loc=12/40.831/-73.917&city=bronx-ny&area=B10&adview=ful...], last accessed 21 October 2023.

Rothstein, R, 2018. *The Color of Law: A Forgotten History of How Our Government Segregated America*. New York: Norton Publishing.

Sanborn Fire Insurance Map from Ogdensburg, St. Lawrence County, New York. Sanborn Map Company, Sep, 1898. Map. Retrieved from the Library of Congress, www.loc.gov/item/sanborn06140_003/

Sanderson, E, 2009. *Manahatta: A Natural History of New York City*. New York: Harry N. Abrams, Inc.

Saunders, P, 2019. "Is Chicago's Legacy of Segregation Causing a Reverse Great Migration?" *Chicago Reader*, Jan 24. https://chicagoreader.com/news-politics/is-chicagos-legacy-of-segregation-causing-a-reverse-great-migration/, last accessed 26 December 2022.

Schiller, P L, and Kenworthy, J, 2017. *An Introduction to Sustainable Transportation: Policy, Planning, and Implementation*, 2nd edition. London: Routledge

Schuetz, J, 2022. *Fixer-Upper: How to Repair America's Broken Housing Systems*. Washington, DC: Brookings Institution Press.

Schwartz, A, 2021. *Housing Policy in the United States*, 4th edition. New Philadelphia, PA: Routledge

Squires, G D (Ed.), 2002. *Urban Sprawl: Causes, Consequences and Policy Responses*. Washington, DC: The Urban Institute Press

Talen, E, 2005. *New Urbanism and American Planning: The Conflict of Cultures*. Philadelphia, PA: Routledge.

US Census, n.d. Generic American Community sSurvey (ACS) data published by the US Census Bureau. Online resource available at census.gov/programs-surveys/acs/.

U.S. Census. *U.S. Census Bureau Construction Spending - Characteristics*. n.d. Online source https://www.census.gov/construction/chars/current.html, 28 July 2023.

US Census, 2000. *Historical National Population Estimates*, July 1, 1900, to July 1, 1999. Sourced from Population Estimates Program, Population Division, US Census Bureau. resource available at https://www2.census.gov/programs-surveys/popest/tables/1900-1980/national/totals/popclockest.txt, last accessed 6 March 2023.

US Census, 2021a. *Historical Population Change Data (1910–2020)*. https://www.census.gov/data/tables/time-series/dec/popchange-data-text.html, last accessed 24 May2023.

US Census, 2021b. *United States Census Bureau today Delivers State Population tools for Congressional Apportionment*. https://www.census.gov/library/stories/2021/04/2020-census-data-release.html#:~:text=The%202020%20Census%20shows%20that,7.4%25%20since%20the%202010%20Census, last accessed 31 May 2023.

USGCRP, n.d. *US Global Change Research Program at North Carolina State University*. Online resource available at https://scenarios.globalchange.gov/loca-viewer/

USHUD, 2023. *US Department of Housing and Urban Development. Housing Developments.* Online resource available at https://hub.arcgis.com/datasets/fedmaps::public-housing-building

Vale, L, and Freemark, Y, 2019. "The Privatization of American Public Housing: Leaving the Poorest of the Poor Behind." Anacker, K, Nguyen, M, and Varady, D (Eds.), *The Routledge Handbook of Housing Policy and Planning*, pp. 189–206. Abingdon, UK: Routledge. doi:10.4324/9781315642338-15.

White, N, and Willensky, E, 2000. *AIA Guide to New York City*, 4th edition. New York: Crown.

Wilkerson, I, 2011. *The Warmth of Other Suns: The Epic Story of America's Great Migration.* New York: Vintage.

# 3 Contemporary Design Adaptations and Policy Interventions

## 3.1 THE CONTEMPORARY HOUSING LANDSCAPE

In Chapter 1, we framed the housing challenges in the United States in three ways – considering the housing supply challenge, the housing affordability challenge, and the lack of housing for the most vulnerable, the homelessness challenge. In this chapter, we expand and complicate these ideas further by discussing the contemporary design adaptations and policy interventions that have emerged recently, that is, in the last three decades. Each adaptation and intervention attempts to address one or more of these challenges, and in doing so, has created new problems for planners and city managers. We return to our socio-behavioral and cultural definitions of housing also referenced earlier to begin this discussion.

We will use the Census Bureau's definition of housing units throughout our discussion of housing. According to the bureau's definition, "a housing unit is a house, an apartment, a mobile home, a group of rooms, or a single room that is occupied (or if vacant, is intended for occupancy) as separate living quarters. Separate living quarters are those in which the occupants live and eat separately from any other persons in the building, and which have direct access from the outside of the building or through a common hall" (U.S. Census Bureau, 2023a). The accepted definition of a housing unit is not related to the entity providing/maintaining the housing unit. A housing unit may contain multiple people that occupy the same unit, as in a family occupying a single-family home. Alternatively, a housing unit can contain only one person, such as a single person occupying a unit in an apartment building. We can infer that a housing unit includes living spaces that are separate and private for occupants and has access to the outside without having to pass through private spaces assigned to other persons. In other words, a residential housing unit is imbued with expectations of privacy accorded by law and societal norms.

About 65% of housing in the United States is in the form of single-family homes.[1] Appropriately, this housing type occupies a prominent place in conversations about housing, especially the production and financing of new homes. Census data suggests that about 10% of housing units are vacant (U.S. Census Bureau, 2023b). There may be many reasons for these housing units to remain unoccupied; for example, some of these vacant units could serve as short-term accommodation, as vacation homes, as temporary rentals (a way for owners to generate additional income), or as second homes that are used seasonally and remain unoccupied for a good part of the year (in some counties of the United States, such vacation homes constitute over

DOI: 10.1201/9780429243066-3

50% of all housing units). Vacant properties could also be part of an inventory of properties listed for sale or rent, or the properties could be in foreclosure proceedings. Vacant residential properties often contribute to negative public perceptions of a neighborhood. Funding for home sales in the United States comes from a variety of sources. As Figure 3.1 (data source, US Census and HUD, 2023) shows, in the early 2000s conventional mortgage loans played a dominant role in financing home purchases. However, with the recession of 2008 there was an overall decrease in home sales, but the share of government-backed loans, such as FHA and VA loans, gained prominence as alternate funding sources. As the economy recovered and the housing market stabilized, conventional loans regained their popularity and the market saw a steady rise in home sales funded through traditional channels until 2020.

The remaining 35% of housing includes multi-family housing (e.g., apartments or condominiums), manufactured homes, and group quarters. When considering total housing stock (the number of newly constructed housing units plus previously built units available for use), it is useful to remember that a certain percentage of units will become obsolete every year (e.g., because of the removal of structurally unsafe units, or the removal of housing units to create non-residential uses). Figure 3.2 (data source, US Census Bureau, 2023b) shows the creation of types of privately owned housing from 1970 to 2020. Single-family homes continue to be the dominant type of housing units being built and buildings with 2–4 units represent a very small number. It should be noted that although the upturn of creation from 2010 to 2020 for both single units and properties with five units or more, the total number of housing units remains on a downward trend.

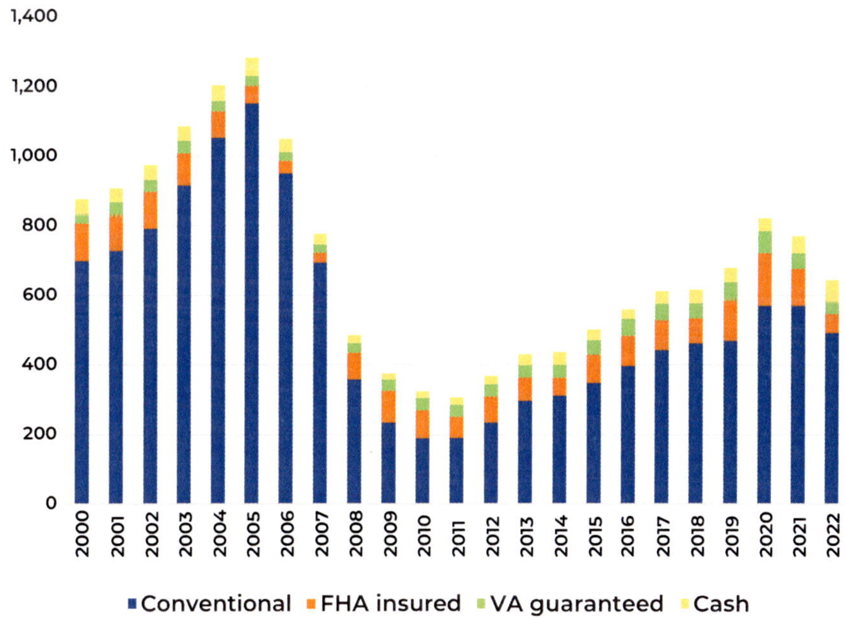


**FIGURE 3.1** Home Sales from 2000 to 2022 by funding source

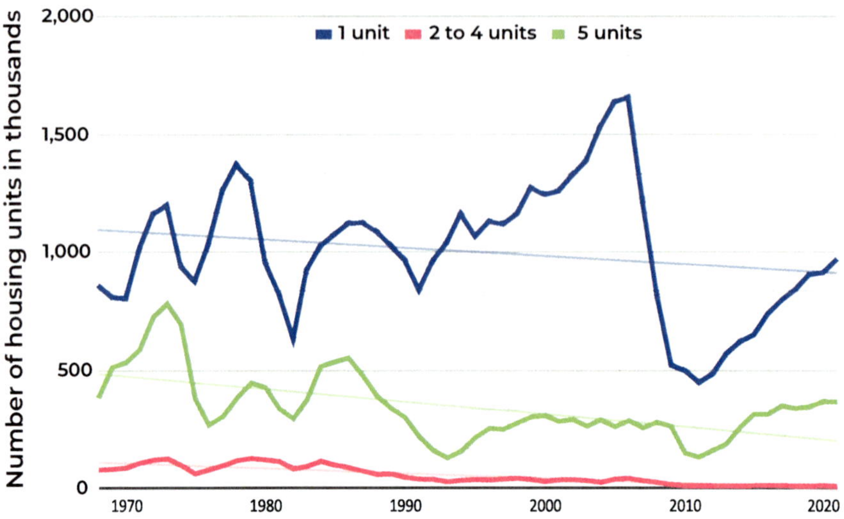


**FIGURE 3.2** New privately owned housing units from 1970 to 2020

Group quarters make up approximately 3% of all housing and these residential housing units are not considered part of the housing unit count in mainstream discussions about housing. Group quarters represent diverse groups of the population. They are defined by the US Census as "places where people live or stay in a group living arrangement that is owned or managed by an organization providing housing and/or services for the residents" (US Census Bureau, 2021). These places include college dormitories, military bases, nursing facilities, group homes, worker homes and prisons. The socio-economic and demographic characteristics of residents living in group quarters can vary widely, depending on why/where unrelated individuals and households are living together. In Figure 3.3 (data source, US Census Bureau, 2021), when we break apart group quarters between institutional and non-institutional, we can see men have a significantly higher presence in institutional settings, likely from prison populations. In Figure 3.4, (data source, US Census Bureau, 2021), which breaks down group quarter residents by age group, we can see that a majority of 18–24-year-olds are in non-institutional settings, likely college dormitories and military bases. The prevalence of group quarters can impact surrounding neighborhoods because they usually offer a range of support services that bring in increased presence of people, cars, and other activity into the neighborhood. Very often, they do not conform to the scale or character of the neighborhood. They often attract protests from NIMBY (Not-In-My-Backyard) groups citing concerns such as increased traffic, overflow of cars parked on residential streets, noise concerns, and security concerns. While these may indeed be valid considerations, opposition based on nonconformity with existing neighborhood character can be thinly veiled prejudice.

Housing is not a single area of specialization, even though it may seem as such to the outsider. Housing specialists in the private sector can include developers, financiers, architects, and realtors. Many more intermediaries are involved when housing production is supported through the use of public funding sources. The sheer

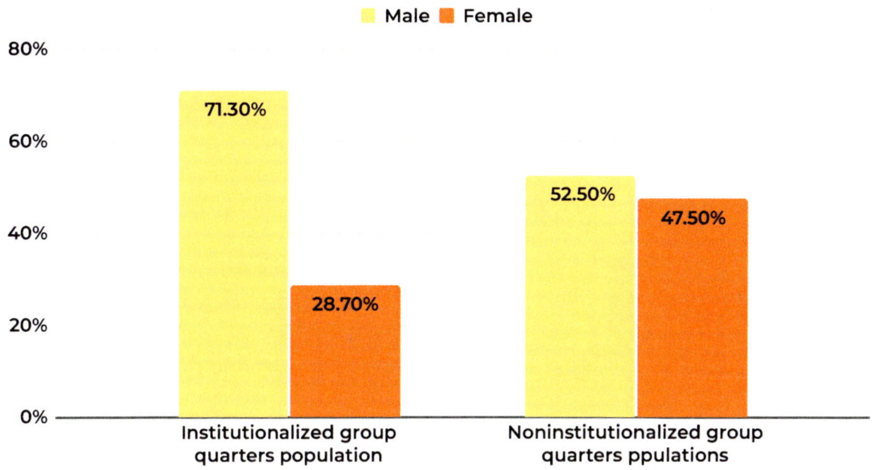


FIGURE 3.3 Group Quarters Demographics: comparative analysis of gender distribution in institutional vs. non-institutional settings

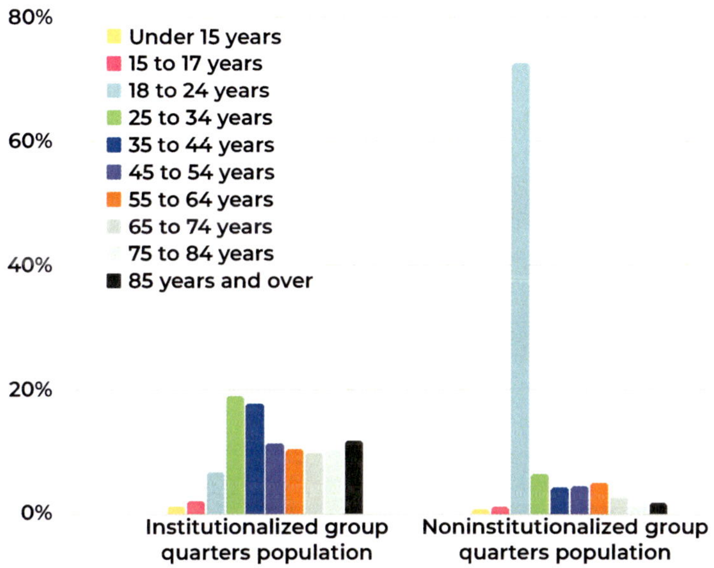


FIGURE 3.4 Comparative analysis of institutionalized vs. non-institutionalized populations

complexity of the enterprise has led to a culture of hyper-specialization. As we encourage housing specialists to explore and take advantage of the wide range of tools and methods available under the umbrella of Geographic Information Science or GIS applications, we note that establishing a common vocabulary becomes very important. For instance, the previously listed housing specialists are likely to have a common understanding of what they mean when they reference "neighborhood amenities". They impose a social values-based assessment of the businesses and services that are likely to be viewed by the general public as "enhancing" the desirability of a neighborhood. For a data analyst who is tasked with mapping or displaying the same neighborhood amenities – the phrase is coded, and they are not able to operationalize the phrase to translate it into a mapping operation. Thus, neighborhood amenities can only be mapped, if additional information and insight are provided, for example, by describing the types of services or businesses that are considered desirable and worthy of being listed as a neighborhood amenity. For example, as necessary as they are, shoe-repair or appliance-repair shops are not considered desirable neighborhood amenities in affluent neighborhoods because of the societal meaning-making that occurs around the concept of repair (suggesting thrift rather than affluence).

Describing a housing development and situating it in the context of the lived experiences of a neighborhood require a classification scheme that does not focus on the architectural design or housing form alone, although form (appearance) is one aspect of a meaningful description. Yet, housing and neighborhood characteristics are strongly influenced by activities (functions) that occur within that housing development, making function a part of that classification. Finally, the number of people present in the development (density) becomes a variable to consider because of its impact on the neighborhood – traffic being an often-cited example, although demands on water supply and sanitation could be considered within this category. GIS specialists opting to work with housing specialists would do well to consider the complexity of classification schema (typologies, in other words) used by housing specialists.

### 3.1.1 Housing Typologies

Housing typologies organize the different types of residential structures, focusing on a range of variables. In architectural terms, a typology may emphasize a design aesthetic, which in turn can also communicate embedded information about a building's height (bulk), the number of rooms (indirectly addressing density). Architects and historic preservationists use terms such as "Cape Cod", "Colonial", "Craftsman", or "Mid-Century Modern" to describe individual properties, focusing on architectural design, the choice of materials, or a cultural characteristic that evolved over a period of time. Realtors may describe the same properties with some additional details, for example, a Colonial with $x$ number of bedrooms and $n$ number of bathrooms. In this book, we will not focus on the design and style of housing because these typologies have evolved over time influenced by availability of materials, methods of construction, and cultural norms. Our research suggests that most neighborhood-level typologies are purpose-built to achieve and accommodate decision-making. For example, historic preservationists may focus on a typology that organizes a neighborhood based on the historic styles of housing and the age (date of construction).

Our discussion of housing typologies focusses on "bulk" (form) and function. In so doing, our framing is closely aligned with the New Urbanist interpretations of housing/neighborhood typologies that emphasize (1) the relationships of the house to the street, (2) the relationship of the street to the neighborhood, and (3) the neighborhood to its location within the city/region. Bulk influences how the structure is experienced at the street scale. By adding function in a consideration of housing typology, we consider the number of people using a particular building type. Density (the number of people within an area) is a computable measure that impacts planning for support services – such as grocery stores or public transportation (we will revisit this topic when we talk about how to deploy GIS for such analyses in Sections 5.5 and 6.2). When contemplating housing typologies, taking into account both density and bulk, we observe the potential for diverse activities and uses. For example, a mid-rise apartment building can consist of individual market rate apartments, supportive housing like a drug rehab facility or serve as campus housing for a university. Figure 3.5 is an infographic that summarizes the housing typologies that we have identified.

We organized the facilities that accommodate residential living by considering both form and function as: (i) single family, (ii) multi-family, and (iii) supported and transient housing. Among these function groupings, single-family units are the dominant type in the United States. We restate this for emphasis: low-rise, detached, and single-family homes on individual lots are the dominant housing type for residential housing throughout the United States. According to the American Community Survey (ACS, 2020, 5-year estimate) nearly 68% of all housing units in the United States are single-family homes and of that, only 6% are attached homes. Figure 3.2 (three pages back) shows historical data for completions of privately owned housing units. The second function type is multi-family dwelling units, which include any residential structure that houses any number of dwelling units larger than one. Multi-family dwelling units can be further segmented to include (i) low-rise detached

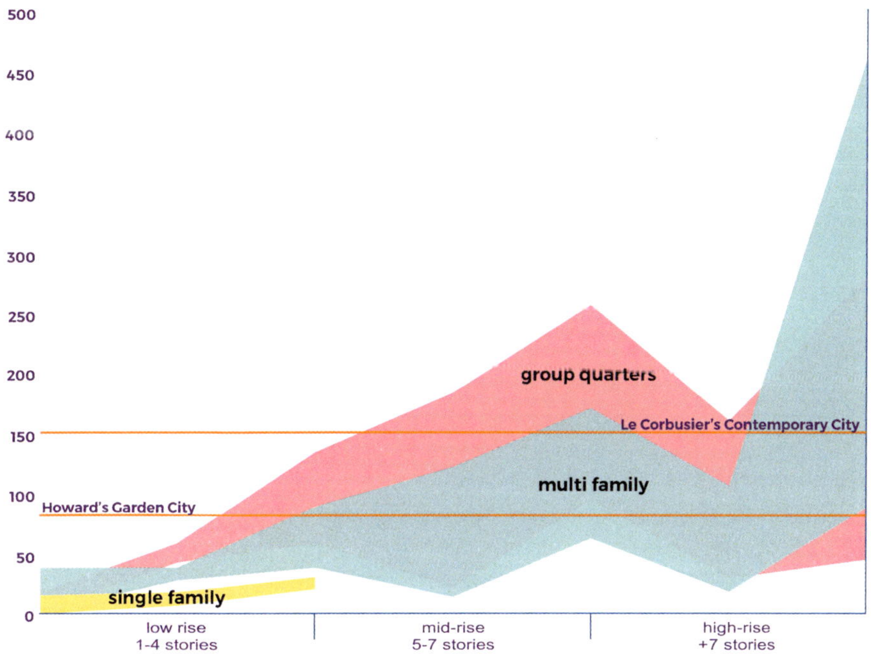


**FIGURE 3.5** Charting housing typologies from low-rise to high-rise, exploring the range of building heights and density

buildings that can accommodate two to four dwelling units, (ii) small apartment complexes accommodating eight to ten dwelling units, and (iii) high-rise structures accommodating hundreds of dwelling units. The multi-family housing type can be found at widely varying densities; in other words, a multi-family house that includes two dwelling units can be situated on a single-acre lot that can also accommodate a medium-rise multi-family dwelling unit with upwards of 200 dwelling units per acre. To a large extent, these differences are governed by local zoning laws.

The New Urbanist movement has directly confronted the tension between the uses (functions) of a building and its appearance (architectural and physical characteristics) (Talen, 2005). The movement's proponents have argued that re-scaling buildings to "fit" the existing styles on a street or neighborhood could help reduce negative reactions against density or certain types of functions such as supportive housing being placed in a neighborhood. Furthermore, the movement's proponents have reasonably argued that increasing density is not an all or nothing proposition. Although the lowest densities are reasonably associated with a single-family dwelling unit, and the highest densities are likewise associated with high-rise multi-family dwelling units, Figure 3.5 illustrates that there is a great deal of variation and possibilities for gradually increasing density in an urban environment.

In the design of multi-family units, it appears that two to four family units have not been popular in recent years. Older cities like Buffalo, New York, or Milwaukee, Wisconsin have a robust supply of duplexes and triplexes because multi-generational families lived together in close proximity and these types of dwelling units also allowed for creating opportunities for rental income. However, the production of these types of units has not risen in the past two decades, even after the great recession of 2008. Missing Middle Housing (Parolek, 2020) is a relatively new movement that advocates for modestly scaled residential buildings with multiple units in walkable neighborhoods. Figure 3.6 begins to identify this "missing middle" bulk and density in NYC when the % of building types is applied to the typology chart. Different architectural forms and massing can increase the density without unduly affecting neighborhood character. Missing Middle housing typologies advocate for a return to duplexes, triplexes, fourplexes, townhouses built as row houses or around a courtyard, and live/workspaces (shops on the street level, house above), as innovative ways to increase densities. When mapped on our housing typology infographic (see Figure 3.6), we can see the missing middle density in NYC. Ultimately, the drive to grow missing middle housing acknowledges that there is a need to move beyond the dichotomy between single-family housing and high-rise apartment housing, regardless of ownership arrangements. It also acknowledges the pre-eminence of low-rise/low-density housing as the preferred option for most Americans.

This discussion of housing typologies should encourage housing advocates and GIS specialists to examine the complex relationships between architectural and urban design methods and their application to public policy approaches addressing the housing challenges we discussed in Chapter 1. On the technical side, the research arms of large firms like Arup Foresight and KPF Urban Interface are developing the tools to capture these complex relationships. And the overall housing shortage and the relative unaffordability of housing in many major markets can (in theory) be addressed by increasing the types of housing that are actually built and made

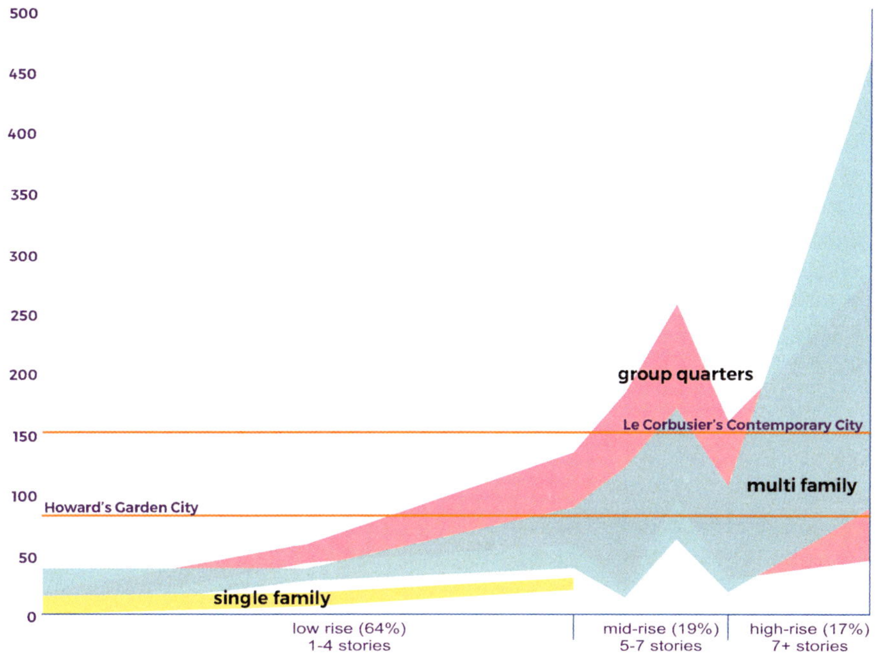


**FIGURE 3.6** Charting housing typologies in New York City

available by the private sector. Undoubtedly, the cost of housing production and returns on investment influence these decisions, but changing legislation about what types of housing is allowed to be built in certain communities must also become part of this conversation. The innovations discussed in Section 3.2 begin to address these concerns.

## 3.2 HOUSING DESIGN INNOVATIONS

While the image of single-family homes on large lots has come to epitomize the American idea of a home/residential living unit, this image is largely a post-World War II ideal (Gans, 1967, 2017). In Chapter 2, we discussed the trends that created and shaped these outcomes. In this section, we discuss a few housing design innovations that have expanded the available range of housing alternatives.

### 3.2.1 Single Room Occupancy (SRO) Units

A Single Room Occupancy Unit or SRO is a residential unit that provides private dwelling quarters with access to shared bathroom and kitchen facilities. Although the affluent lived in residential hotels (a type of SRO) as early as the 1800s, this was not the norm. Conventionally, SRO residents rented/leased small spaces (minimum 120 square feet), and had a safe space to rest, store their belongings, and a permanent address for an extended period of time. In smaller towns, SRO options were provided by homeowners who rented out rooms in their house (taking in boarders). In other instances, a rooming house/boarding house (several rooms available for rent in a single building) offered a similar option for a group of unrelated individuals. Throughout the early 20th century, in cities such as New York, Chicago, and San Francisco, SROs provided affordable shelter options for single men and women who were under-employed or working low-wage jobs, see Figure 3.7 (Byron Collection). Although they provided for basic needs and may have prevented these individuals from becoming homeless, they were not viewed as a desirable option because of unhealthy and unsanitary living conditions. Consequently, SRO housing stock was demolished or zoned out of existence citing health and safety concerns.

Post-2000, SRO housing has been rediscovered as a viable housing option for transient and hard-to-house populations, including those populations who are in recovery. Modern SRO units include in-unit bathrooms and kitchens, see Figure 3.8 (SRO Housing, 2023). These SROs are directly or indirectly supported by the State or philanthropic organizations. As a housing type, college dormitories, retirement homes, and long-term care facilities have the physical characteristics of SROs although they are not classified as such. The main distinction appears to be the type of ownership and the social class of people residing in these units (US Department of Housing and Urban Development (HUD), 2001).

Cities such as San Francisco have introduced measures to stabilize and protect existing SRO housing stock and amend restrictive zoning laws that prevent new SRO

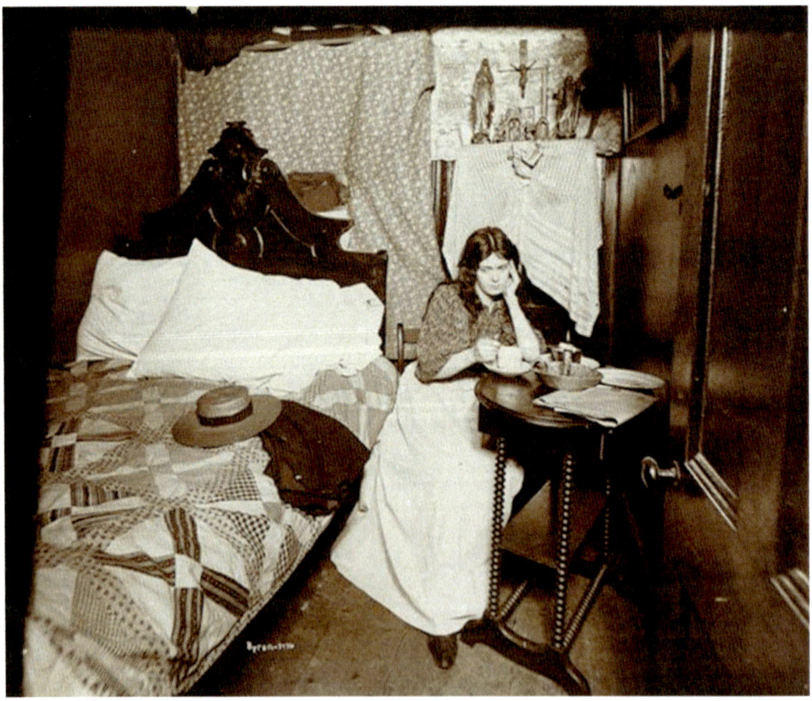

**FIGURE 3.7** SRO historical photo

FIGURE 3.8 SRO contemporary photo

housing from being constructed. In Miami-Dade county, SROs run by private companies use a Section 8 Single Room Occupancy Program for very low-income people that are on the street or in a shelter. While far from an ideal option, SRO housing units, especially those combined with supportive services, are a reasonable affordable housing option for low-income people. However, zoning policies may not allow the production of new SRO units and exclude SRO housing from the range of housing typologies that are available in many cities.

### 3.2.2 Accessory Dwelling Units (ADUs)

Accessory Dwelling Units or ADUs have been known over the years by a number of names, such as in-law units, granny flats, secondary units, or mother-daughters. They refer to a part of a residential property that is shared and can be used for the purpose of renting to help the owner recoup the costs associated with purchasing and maintaining the property. Typically, these units are located in a single-family residence, such as a basement, attic, or a garage. Although these ADUs have been present over a period of time, many building codes and zoning regulations have systematically prohibited their use, usually citing health and safety concerns. With ADUs, we can distinguish between those that can be attached to the primary residence and those that are detached from the primary residence.

Many cities endeavor to formalize and legalize the existence of ADUs by amending zoning laws and ancillary regulations. At a time where the square footage of

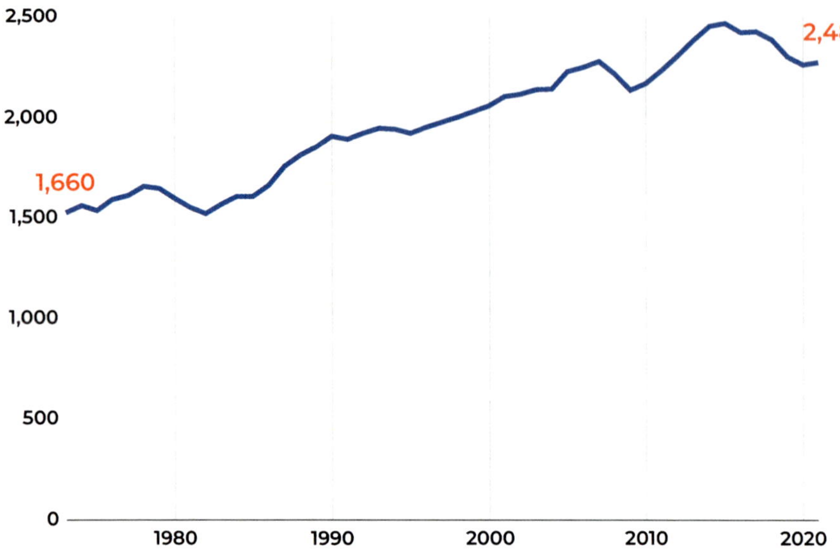


**FIGURE 3.9** Increase in square footage of single-family homes

single-family homes has increased, see Figure 3.9 (data source, US Census, 2023d), there is even more opportunity than before to create these ADUs in residences. ADUs offer low-income homeowners a practical way to lower the burden of housing costs and encourage and support property ownership. Some proposed regulations limit ADU uses such as encouraging the creation of multi-generational households by restricting leasing only to family members.

Allowing ADUs generally requires a municipality to make changes to their zoning regulations. This requirement has contributed to the difficult legal implementation of ADUs in many places. Existing regulations governing parking, allowing accessory buildings on lots, and single-family zoning place limits on scaling up the use of ADUs as a viable housing option. For example, parking regulations in many ordinances require a new parking spot be created for a new dwelling unit. This regulation makes the addition of an ADU more costly and less feasible. In California, the state legislature made sweeping changes to ADUs, allowing them to be built in areas zoned for single-family housing. The government code was amended so that a city does not require replacement of parking if a garage was converted for the ADU and waives the need for parking if the ADU is within a half mile of public transportation, if it is in a historically significant area, or if the ADU is part of the primary home or accessory structure (California Department of Housing and Urban Development, 2022). California also asked all municipalities to develop a plan on how to adopt ADUs within their cities and counties in order to promote a statewide effort to increase ADUs. The California Health and Safety Code (HSC) Section 65583 (c)(7) requires that cities and counties develop a plan that incentivizes and promotes the creation of ADUs that can be offered at affordable rent for very low to moderate-income households. These new regulations' effectiveness can be seen in the overall increase in ADUs in California as depicted in Figure 3.10 (data source, ADU 2022).

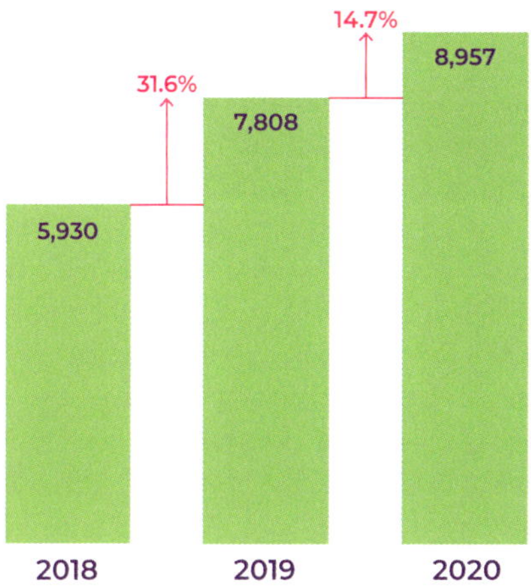


**FIGURE 3.10** Unlocking Housing Potential: ADU creation in California from 2018 to 2020

ADUs have been written into the zoning ordinances of places like Lexington, Massachusetts, Santa Cruz, California, Portland, Oregon and Fauquier County, Virginia. These places have removed restrictions in the zoning for the allowance of accessory buildings that house ADUs and the restrictive single-family zoning. In Lexington, Massachusetts, the ADU code section allows for attached ADUs for lot sizes up to 10,000 square feet but allows for a detached ADU on lots that are at least 18,000 square feet.

### 3.2.3 Manufactured Housing

Manufactured homes were traditionally called "mobile homes". Mobile homes, as defined by the Department of Housing and Urban Development (HUD) do not require a building permit, have no foundation, and are built to have a gear that allows them to be moved on their own chassis. Mobile homes are "manufactured" in a factory and then moved or placed on an available plot of land. The land can be a single parcel, or part of a trailer park that houses many manufactured homes. Manufactured homes began life as a home on a trailer that was pulled by an automobile. In the 1930s these trailer homes were typically used for auto-camping. After WWII trailers began to be used for housing. These homes were used for temporary accommodations for migrant workers, and for use in neighborhoods and communities that are affected by natural disasters like floods or tornados that destroyed existing housing stock. However, as housing affordability has decreased, manufactured mobile homes, formerly used for temporary housing are now used as a permanent housing solution.

Manufactured mobile homes are regulated by HUD, and since 1976 all manufactured homes must meet certain standards and be given a sticker from HUD that certifies the home. The federal Manufactured Home Construction and Safety Standards Code (the HUD Code) requires compliance for fire resistance, energy efficiency, strength, and durability. Some jurisdictions require HUD compliance for a home to be located in a trailer park and use the HUD certification to avoid imposing additional code requirements. The Manufactured Housing Institute (MHI), an industry group, cited HUD data to state that 22 million people live in over 8 million manufactured homes in the United States (Manufactured Housing Institute MHI, 2023).

Manufactured home production is completely based in the United States, with Texas leading the production and consumption of manufactured homes (US Department of Housing and Urban Development, 2023). According to the ACS 2020 (5-Year Estimates), in states like New Mexico and South Carolina, mobile units make up 16% of all housing unit types, unlike states like Nebraska and Utah where only 3% of all housing unit types are designated as mobile homes. There is a high variability in the percentage of residential dwelling units that are designated as mobile manufactured homes and there is intra-state variability as well. Table 3.1 (data source, US Census, 2023c) shows nine counties with over 50% of the housing share as manufactured homes. Even though Utah has only 3% of all housing units as manufactured homes, one county on the list is from Utah. Geography matters, as does public acceptance of manufactured homes because living in a manufactured home continues to be stigmatized (U.S. Census Bureau, 2023c).

Manufactured homes have a number of options for location and ownership. In itself, a manufactured home is in-fact movable and typically not considered "real" property, but rather as personal property. This personal property can be located in a trailer park, on an owned parcel of land, or as part of a cooperative, where mobile home owners also own a share of the mobile home park. Formerly categorized as "temporary use", such locations did not have the privileges of residentially zoned areas, meaning that the residential status was tenuous. The status of the mobile home changes from being personal property to real property when the mobile home is located on a parcel of land that has the same owner. When a manufactured home is located in a trailer park, the manufactured homeowner is in fact renting a parcel of land within a community. There is risk associated with this type of ownership because the trailer park owner can evict the tenant. The eviction can happen if the trailer park closes or sells the land. Although a manufactured home is by definition able to be moved, it is a difficult and costly endeavor.

**TABLE 3.1**
**Counties With Over 50% Manufactured Homes**

| County | State | % Manufactured Homes |
|---|---|---|
| Daggett | UT | 50.2 |
| Glades | FL | 50.7 |
| Brantley | GA | 51.6 |
| Gilchrist | FL | 51.9 |
| Suwannee | FL | 52.3 |
| Quitman | GA | 54.4 |
| La Paz | AZ | 55.1 |
| Echols | GA | 55.9 |
| Lander | NV | 56.9 |

Manufactured homes as a part of housing cooperatives are another innovative way to build wealth. Homeowners in trailer parks that are part of a cooperative can own a share of the land and therefore mitigate the risks associated with renting the land. Shared ownership also stabilizes the community as there is an investment in the land as well as the manufactured home. These types of cooperatives have been occurring within senior retirement communities allowing for a low-cost way of living, but at the same time protecting wealth. The New Hampshire Community Loan Fund is one of the oldest lending programs that serves 146 resident-owned manufactured home communities (ROCs) in New Hampshire (Community Loan Fund, 2023). The Loan Fund provides the infrastructure, technical assistance, and training to create and support manufactured housing cooperatives. There are additional financing options available to purchase manufactured homes. The HUD FHA program insures mortgages for manufactured homes sold with land, which is a Title II loan. The Title II program allows for a loan when land is not owned. In addition to the FHA loan programs, manufactured homes can be financed as personal loans since they are considered personal property.

Manufactured homes allow for the American Dream of home ownership to expand to people with a lower income that may not be able to afford a home built with traditional materials. However, as seen in Figure 3.11 (data source, US Census, 2017) the share of manufactured homes has been decreasing since 2009. The Biden administration considers investments in manufactured homes as one of the strategies to address the housing crisis. HUD expanded the Title I guidelines for manufactured homes and incorporated them into the Single Family Housing Policy Handbook 4000.1. The move aims to "enhance value determinations, expand allowable income sources, and allow additional flexibility in calculating student loan debt". These new revised guidelines are aimed at combating the housing crisis and HUD has identified these manufactured homes as a key opportunity of doing so.

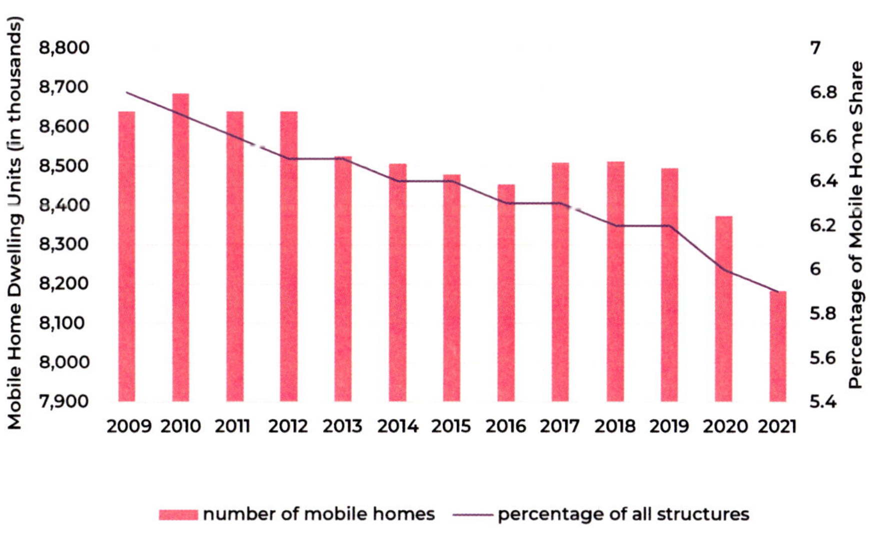


**FIGURE 3.11** Declining trend for mobile home dwelling units in the United States

### 3.2.4 Tiny Homes

Tiny homes are just what they seem: small homes that are typically between 60 and 400 square feet. The average size of a single family in the United States ranged from 2,473 sf in 2020 to 2,485 sf in 2022 making these tiny homes significantly smaller than the average single-family home, see Figure 3.12. Tiny Homes are also smaller than their manufactured home counterpart, whose average is 1,184 sf. According to the Tiny Home Society, an intentional advocacy group that advances the concept, tiny homes can include houses on foundations, houses on wheels, accessory dwelling units (discussed earlier in this chapter), and park model recreational vehicles or RVs. The tiny house movement "offers more affordable and sustainable housing alternatives for millennials, environmentalists, and others seeking unconventional living" (Alexander, 2022).

Tiny homes that are built on trailers are typically coded and regulated as recreational vehicles (RVs). They can be certified as a homemade RV, but often are not. The Department of Motor Vehicles (DMV) inspects the trailer that supports the home. There are issues with applying the building code to tiny homes, as they do not meet many of the regulations as set forth in the contemporary codes for residential living, such as minimum widths for rooms and egress requirements. Right now, tiny homes are not considered to be permanently occupied dwellings and rather a place where people camp temporarily, but as more Americans move into these tiny homes on wheels for permanent living, building codes will need to be updated to ensure safety.

Although many environmentalists praise the limited impact that tiny houses have on the environment there are a number of factors to consider their efficiency. As discussed, existing building codes do not apply to these structures and therefore the energy code requirements for traditional homes are not implemented for these homes even if they are constructed with similar materials.[2] There is a high exterior surface area compared to the interior space. The study by Mukhopadhhyay et al.

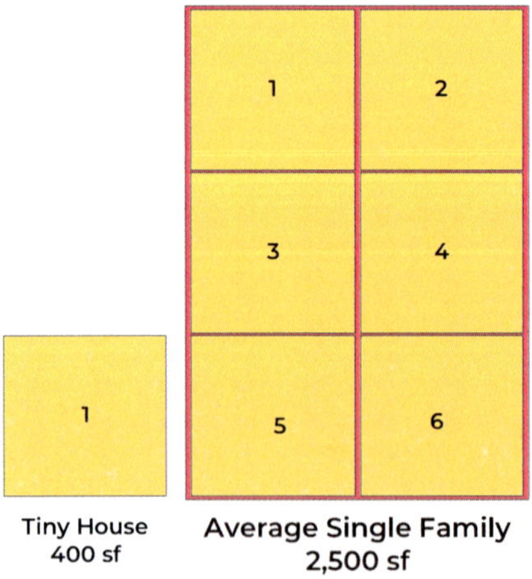


FIGURE 3.12 Scaling Down: size comparison of tiny homes to the average single-family home in the United States

(2019) conducted in a cold climate found that air filtration rates did not comply with building code standards. Heating these homes comfortably was also observed to be a challenge. These issues may be further addressed when tiny home building codes propose alternative standards and guidance to address these problems.

In addition to emphasizing individualism and a boutique lifestyle, tiny homes' proponents are advocating for their use to address the challenge of homelessness. As the body of regulations regarding tiny homes is still underdeveloped, there is as of now an additional ethical burden for developers to assure sites to locate a group of homes are appropriate and safe, and a relatively dense settlement of homes can be built and sustained. Highway underpasses, vacant lots, and other under-utilized locations can serve as safe shelter options, in essence creating humane dwelling conditions to replace homeless encampments. However, addressing hygiene, sanitation, health, and safety will require formal guidelines for the creation and maintenance of these housing alternatives.

### 3.2.5 Cohousing

Cohousing is a concept originating in Denmark and introduced to US audiences by McCammant and Durret (2011) that blends private living with shared open spaces and services, managed through cooperative principles. Cohousing includes attention to: (i) socio-cultural characteristics such as ensuring participatory processes in designing and managing the community, (ii) design characteristics that facilitate community interaction and engagement, and (iii) governance characteristics that consciously de-emphasize hierarchies and emphasize communitarianism. Kibbutz, for example, can be considered an agrarian co-housing model, embedded with the religious and cultural context of Israeli society. The cohousing model can support intentional communities such as cooperatives, planned unit developments, and retirement communities.

Elder Cohousing as a form of retirement housing can make housing more affordable by promoting the sharing of common areas such as cooking facilities, and resources such as on-site health care services to reduce overall expenditures for individuals. Communal living by design requires zoning variances and societal acceptance. While a conventional retirement community can offer a similar lifestyle, cohousing arrangements are defined by a culture of cooperation and collective responsibility for the wellbeing of the community. Figure 3.13 (Google, 2023) is an aerial view of an elder care cohousing community in Abingdon, Virginia, where the density of the small community is greater than the surrounding residential neighborhood.

Retrofit cohousing transforms existing suburban layouts to create shared common areas like gardens, passive recreational spaces, and workspaces. By removing fences between backyards, larger centralized and safe open spaces can be shared among six to twelve dwelling units. Likewise, larger houses can be converted to a central dining/kitchen area or club house to serve all the households aligned with the community. Angela Sanguinetti (2015) examined cohousing community locations to better understand the relationships between education levels, political affiliations, and preferences for cohousing alternatives to consider how to diversify cohousing and promote its value outside of a niche of relatively affluent, educated, and predominantly White populations.

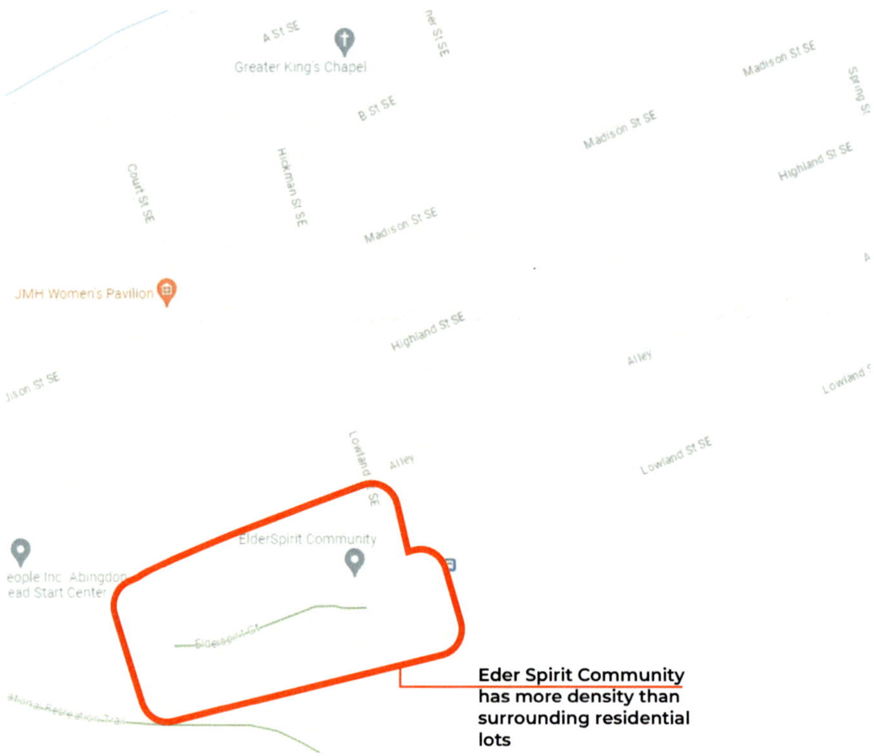


**FIGURE 3.13** Increased density at the Eder Spirit Cohousing Development

### 3.2.6 Transit-Oriented Development (TODs)

Transit Oriented Development (TOD) emphasizes the creation of intensive high-density development around transit nodes such as light rail stops or train stations. It encourages walking and biking but provides a way to extend regional connectivity outside walkable/bikeable neighborhoods through transit networks. Initially promoted by Peter Calthorpe (e.g., Calthorpe and Fulton, 2001), the resemblance to Howard's Garden City concept is obvious. By integrating the transit infrastructure as part of the densification of development, TODs spur and direct the creation of mixed-income and mixed-use developments. The federal government views TOD as a catalyst to encourage increased ridership for transit systems, improvements in air quality, reduced traffic congestion on the roads, and other environmental benefits. It is also viewed as one way to increase housing affordability and promote neighborhood revitalization (Federal Transit Administration, 2023).

TODs are a practical way to transform the existing suburban landscape that is heavily auto-dependent by increasing densities in transit-rich areas. Zoning changes, removal of parking minimums for new developments, and relaxing other restrictions on height and bulk are essential to the success of TODs.

### 3.2.7 Master Planned Communities

One of the more distinct features of the American housing landscape is the master planned community. These communities are large scale residential developments that are developed like small cities. They are privately developed with financial success in mind. Early examples of master planning communities include Radburn (1929) designed by Clarence Stein and Henry Wright influenced by Howard's Garden City concept from two decades earlier. Other examples include Greenbelt, Maryland (1935) that emerged as a result of the Greenbelt Towns Program conceptualized by Rexford Tugwell and advanced by the federal government to create model communities from scratch for low- and moderate-income people. Part of the New Deal, Greenbelt, Maryland, incorporated design and planning ideas from the Garden City and Radburn.

Reston, located in Fairfax, Virginia, was the brainchild of Robert Simon, Jr., a New York real estate developer. In the early 1960s, Mr. Simon invested in a large swath of land in Fairfax County and envisioned and developed a complete community, including commercial and residential uses. It would also feature different types of housing, including condominiums, apartments, townhouses, and single-family homes. This planning was a departure from the suburban developments that Simon observed on Long Island, where residential zones were largely separate from commercial and business districts. In a speech given in 1965, Simon posited three priorities for the new community. He wanted people to live and work in Reston, with opportunities for both mind and body. He wanted it to be possible for a person to be born and live in Reston until they died. He added that the importance and dignity of each individual should be considered over the importance of the community. With these goals in mind, Reston was laid out and zones for housing, offices, medical, and government areas built around an urban core that included businesses and housing.

Developed at the same time as Reston, the Columbia Metropolitan Planned Community was built by James Rouse in the Washington-Baltimore corridor in the 1960s. Rouse used the money he made selling Carnegie Hall to NYC to fund the development. Rouse incorporated a vision of social planning that was different from other MPCs including opportunities to live and work in the same location, environmental protection goals, investments in schools, parks, playgrounds, and other family-friendly amenities, and the deliberate creation of housing alternatives to serve different income levels. Figure 3.14 (Google, 2023) of Wilde Lake in Columbia demonstrates the walkability that was planned into the small towns that make up the larger district of Columbia. Decisions that were made during its development have been largely successful for racial integration. Today, Columbia is racially diverse with 51.1% of the population of appr. 100,00 residents being non-White. Within the non-White population, 28% are Black alone, 13.3% are Asian alone, and 7.3% are two or more races.

Celebration in Florida was developed by the Walt Disney Company in the early 1990s. This MPC creates a planned town of just under 10,000 residents in 2020, with many walking trails and child-friendly environments. Like Radburn, Celebration is not operated by a public governing organization, but rather a private one. This may be a contributing factor to the lack of diversity in this town. Although the intent to create a racially and ethnically diverse community was articulated early on, the current makeup of Celebration has largely failed to meet those expectations with 76.5% of the population being White. Celebration has only 0.5% Black residents compared to the state of Florida where that number is 17%. The median income of $92,110 in Celebration is also higher compared to Florida's $61,777. Figure 3.15 (Google, 2023) shows the walkability of the town and its central business district that serves a surrounding residential population.

**Blue shade shows 15 min walking distance. From residential areas you are able to walk to community center, school, green paths and parks, and commercial areas (yellow shade)**

**FIGURE 3.14** Walkability of Wilde Lake, one of Columbia, MD self contained villages

### 3.2.8 Universal Design

Universal Design or Inclusive Design principles call for "structures and spaces to accommodate a variety of abilities; be easy and intuitive to use; communicate necessary information, regardless of sensory abilities; minimize opportunity for error; and be able to accommodate different body sizes, postures and mobility" (Institute for Human Centered Design, cited by Lowenkron, 2021). Universal Design will serve elderly, neuro-divergent individuals, children, individuals with physical disabilities, and more. The core concept is that by creating built environments that respond to the needs of our most vulnerable, we can better serve ALL people more effectively. Consider, for example, a house that allows residents to safely age in place – this would require considerations for limited mobilities and reduced sensory perception, not to mention the need for the interior spaces to accommodate wheelchairs or other assistive devices. The initial investments will serve children and adults well, while increasing alternatives for aging adults to remain in the familiar surroundings of their own home. Specific attention to creating housing alternatives to serve aging populations is necessary given the graying of America as discussed in Chapter 1 (see Figure 1.19).

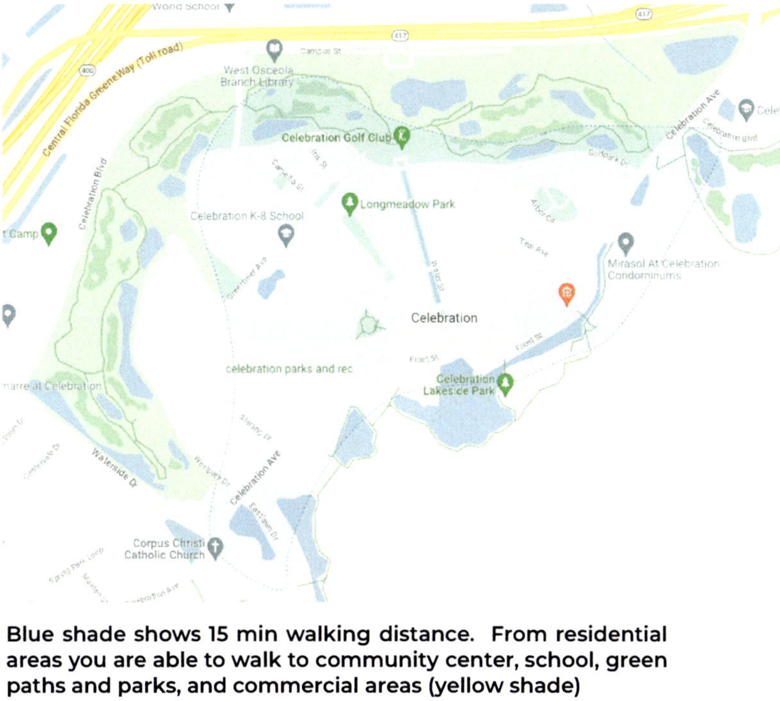


**FIGURE 3.15** Walkability of Celebration, Florida

Collectively, the housing design innovations discussed in this section address ways in which communities can use different approaches to create housing alternatives. Geography, climate, lifestyles, and regional variations in policies and laws all influence the growth and sustainability of specific innovations.

## 3.3 HOUSING POLICY INNOVATIONS

While architects, real-estate developers, urban designers, and entrepreneurs have collectively engaged with developing built environment alternatives to address housing supply and affordability, housing policy advocates, federal, state, and city policymakers have proposed policy innovations and interventions to address different aspects of the housing crisis.

### 3.3.1 RAD Conversion

RAD conversion, HUD's Rental Assistance Demonstration Program, was enacted in 2012. RAD projects are public housing developments that can convert to being managed and supported through private funding sources, while maintaining public ownership. RAD was set up in part to eliminate Public Housing Authorities (PHA) and creates a new entity that could provide better services for the tenants, such as much needed repairs. Deferred maintenance is directly addressed, and all units are converted from Section 9 Public Housing to Section 8 housing. In Section 9 Public Housing residents pay up to 30% of their income, but when converted to Section 8 they pay 30% of their income automatically. A few concerns of the RAD conversion are the longevity of the impact. The immediate impact of RAD conversions is that long-deferred maintenance needs are addressed and the building is restored to its original condition. However, the continued maintenance after the initial repairs are completed is somewhat less assured. Private management companies operate the housing project, and some may do a better job than others. During the first 10 years of its existence, 169,360 units were converted to long term Section 8 housing, including 1,614 different conversions across the country with the median conversion size being 72 units (US Housing and Urban Development, HUD, 2023).

### 3.3.2 Inclusionary Housing/Zoning

Inclusionary Housing or also known as Inclusionary Zoning is an umbrella term for policies aimed at increasing housing availability for low- and moderate-income households. Inclusionary housing can either be mandated or achieved through incentives. As Calavita and Grimes (1998) observed, the impetus of inclusionary housing's creation was the need to solve a spatial problem of concentrated poverty. The original idea purposefully disperses low- or moderate-income households among market rate households in order to integrate diversity of income and race into the segregated populations of American cities. Maryland, California, and New Jersey were among the first states to begin incorporating inclusionary housing. As of 2022, cities in at least 20 states have implemented some form of inclusionary housing.

However, there are some states that have encountered barriers to incorporating this policy into the zoning law because of conflict with a state's ban on rent control. Inclusionary housing obligates the developer to control the rent of a certain number of units in a project and therefore, from a strictly legalistic perspective, developers are creating rent-controlled units, which would be illegal in that particular state. This is true, for example, in North Carolina. There are other states that explicitly ban cities from enacting local inclusionary housing laws as a mandatory action, but generally the incentivized policies are allowed. Once inclusionary zoning policies are established, GIS tools can support the selection of sites (parcels), where new inclusionary zoning policies can be applied to spur housing production, e.g., as part of transit-oriented developments. Figure 3.16 (NYCDCP, n.d.) visualizes all inclusionary zoning locations in New York City.

### 3.3.3 LEED®-ND™

LEED®-ND™ which stands for *Leadership in Energy and Environmental Design for Neighborhood Development* is a rating system developed by the United States Green Building Council (GBC) for "identifying, implementing, and measuring green building and neighborhood design, construction, operations, and maintenance" (LEED, 2023). The GBC has a number of LEED programs that are related to the building scale, such as Building Design and Construction (BD+C) or Building Operations and Maintenance (O+M). LEED ND applies the goals of sustainability to the neighborhood scale. The scale defined for LEED ND is at least two habitable buildings and no larger than 1,500 acres. Residential and nonresidential buildings, as well as buildings that combine both residential and nonresidential uses are eligible for a LEED ND designation. The LEED ND rating measures these goals assigning points or credits to specific themes/considerations. These themes include solar orientation, transportation demand management, mixed use neighborhoods, smart location, local food production, neighborhood schools, compact development, heat island reduction, and tree-lined and shaded streetscapes. For each of these themes, a

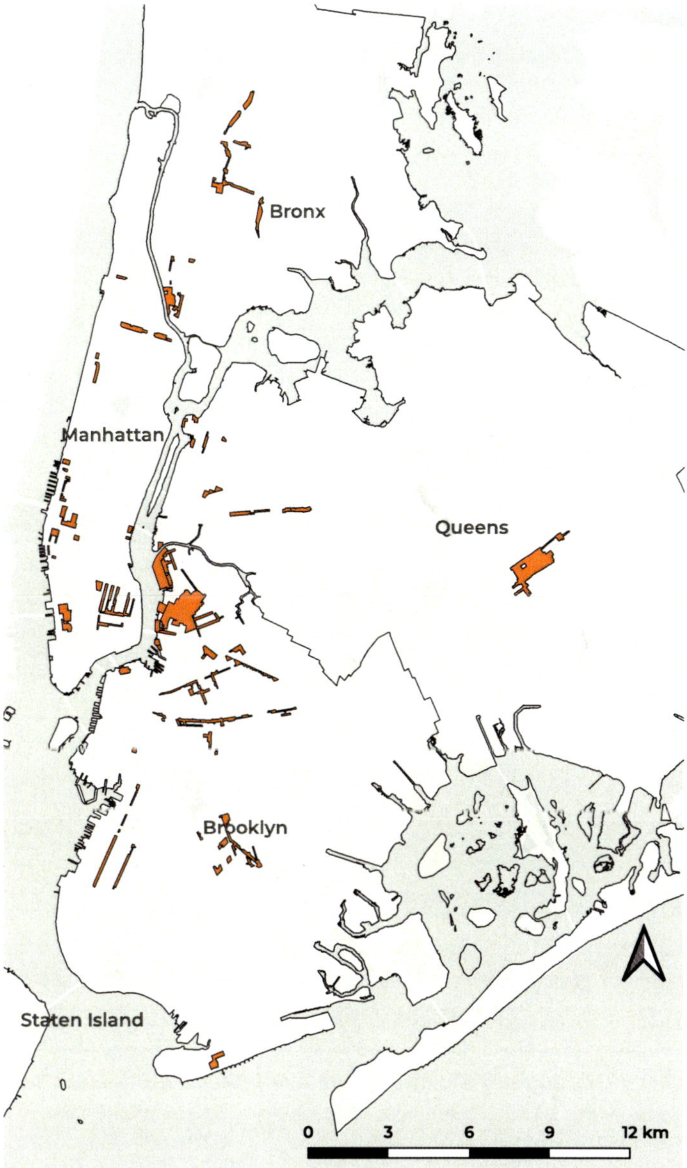


**FIGURE 3.16** Inclusionary zoning developments in New York City

detailed accountability metric including a specific time frame to ensure full compliance is established. For example, under the local food production theme, the building may commit to a neighborhood garden to be constructed by the time the first building is ready for occupancy and be required for the garden to be maintained for 5 years after buildout. The size of the garden and requirements for maintaining it generate a score. In this example, more points will be accrued spending on the space assigned as growing areas and the period of time for which the area will be maintained.

Proximity to jobs and housing as well as housing types and affordability are additional scoring items. For housing types and affordability, credit is given when a project is located in a high-priority redevelopment area, such as a site listed by the EPA National priorities list, a federal empowerment zone, a federal enterprise community site, a federal renewal site, a Community Development Financial Institutions Fund (CDFIF) Qualified Low-Income Community, a HUD Qualified Census Tract, or a designated Difficult Development Area. Another way to gain credit for housing types and affordability is to include a variety of housing sizes and types in the project. For this category, the Simpson Diversity Index is used to score developments. The GBC identifies 20 housing categories, and the Simpson Diversity Index gives a higher score when there is a mix of the types. The list of housing types includes Accessory Dwelling Units.

For Housing and Jobs Proximity, credit is given when 30% of the project's residential total building floor area is located within a ½ mile walking distance of existing full-time equivalent jobs. Another way to gain credit for Housing and Jobs Proximity is to include a nonresidential component on an infill site which is a ½ mile walking distance of an existing rail transit, ferry, or tram stop and within a ½ mile walking distance of existing dwelling units. The scores for the different categories and commitments are combined to meet a total score which establishes the LEED ND Certification of Silver, Gold, or Platinum ratings.

Using GIS, Smith and Bereitschaft (2016) examined light intensity and impervious surface data for the LEED-ND projects and concluded that "by incorporating LEED®-ND™ standards into their land use planning efforts, urban planners may be able to substantially increase the overall sustainability of their urban development projects". The disadvantages of the LEED-ND rating system are that the entire system is voluntary, incentivizing desirable planning and design goals. There is also a concern that developers may invest in commitments that are focused on the physical characteristics (like energy efficiency) rather than the more socially responsible commitments like housing affordability especially over an extended period of time.

### 3.3.4 Subsidies for Energy Efficient Housing

The federal government provides subsidies, commonly called a "green tax credit", for different types of projects that use sustainable energy. These types of credits were first introduced under the Energy Policy Act of 2005, which allowed tax credits for homeowners, builders, and producers of manufactured homes. The Act of 2005 created new federal standards for the energy efficiency of residential and commercial properties. In addition, it offered tax credits for the installation of certain products. The amount of credit varied, but the top credit was $2,000 to builders. New construction and existing buildings were both eligible for the tax credit and the eligible categories for the credit include very efficient HVAC systems, lighting, exterior envelope efficiency, insulated windows, hot water heaters, energy-efficient appliances, and fuel cell installation. The Act of 2005 also set up provisions to create a public housing energy office at HUD (Nadel et al., 2005). The HUD provisions in this act also required public housing to purchase Energy Star equipment. The tax credits for the Act of 2005 covered the years 2006 and 2007.

The American Recovery and Reinvestment Act of 2009 allowed homeowners and builders tax credits from 2009 to 2017. This tax credit program worked similarly to provide tax credits for efficient houses and products. In addition to tax credits, the act funded public housing improvements, improvements for housing of service members, increases to energy efficiency in low-income housing, rehabilitation of Native American housing, and emergency food and shelter for the homeless. Solar equipment credits were also included in this act. Since the end of the credit from Act 2009, the federal government has continued tax credits for solar generation under other programs.

These programs have not been as effective as hoped. The Energy Policy Act of 2005 set a goal of reducing energy use in new homes by 30% by 2015, but this goal, which is hard to measure, was likely not achieved. Additionally, it is worth considering that energy codes and standards have continued to evolve and improve since the Energy Policy Act of 2005. Subsequent revisions to energy building codes have further influenced the energy efficiency of new homes in the years following the act's implementation. The lack of success of these programs also can be attributed to the fact that the programs have mainly focused on new homes, but the vast majority of homes in the United States are existing homes. Subsidies focused on existing buildings can help homeowners and renters save on their energy bills and offset the cost of energy efficient upgrades, making them more affordable. They are a valuable tool that can help make energy efficiency more affordable and accessible to all, if implemented correctly.

### 3.3.5 LIHTC

LIHTC references the Low Income Housing Tax Credit that started in 1987 to provide tax credits as incentives to private investors to encourage them to build or rehab low-income housing. This is a federal program that works hand in hand with individual states to designate eligibility criteria and designate the period of time when low-income housing will be available in the development. LIHTC has produced 3.44 million housing units as of 2020 (US Department of Housing and Urban Development HUD, 2022) and can be regarded as the main privately funded approach for creating low-income housing in the United States. From the perspective of the sheer number of units created, LIHTC is the most successful housing program in US history.

LIHTC units must generally meet affordable rent eligibility requirements that are based on household income as a percentage of the area median income (AMI). The calculation of the percentage of a development that is required to remain below a specific AMI level has been adjusted through the years to allow for more flexibility for builders, as well as some flexibility for renters whose income may have increased over the years but still wish to remain in the same unit. Developers are required to maintain the composition of the AMI that is established for a minimum of 30 years. Exact lengths can be more restrictive on a state-by-state basis.

Unlike a tax deduction, which only reduces taxable income, the LIHTC credits offset dollar-for-dollar a party's tax liability. Developers sell the right to use these credits to investors who want to reduce their federal taxes. The investor's payment for such right, its "capital contribution" to the project, reduces the developer's need to use other financing. This then reduces the developer's debt-service costs, allowing the development to be financially appealing even with below-market rental income. This formula has been successful in attracting private dollars to create affordable housing. LIHTC can also be used to preserve projects funded or supported with other affordable housing programs, including, for example Federal Housing Act (42 U.S.C. § 1437f) Sections 8 (Rental Voucher Program), 236 (Rental Assistance Program), 221(d)(3) (Rent Supplement Program), 202 (for elderly households), 515 (for rural renters), and 514/516 (for farm workers).

Although the credit was authorized by federal law, and reduces federal tax liability, the federal government has put the administration of the program in the hands of the states. Each state has created a housing finance authority (HFA) that allocates credits to developers, administers the state's criteria and bidding process for projects, and monitors developer compliance with program regulations. In Figure 3.17 (USHUD, n.d.), we see municipalities that choose not to utilize the LIHTC program and may face limitations in their ability to create and preserve affordable housing options for their residents.

The builder receives tax credits of 9% or 4% depending on the make-up of the specific project. The tax credit is applied for the established length of affordability. Qualified Allocation Plans (QAPs) are structured to award more tax credit points for specific features like increased time period that the units remain affordable, historic preservation projects, promoting mixed income developments in a low-poverty area or meeting green building standards (Scally et al., 2018). Green building standards are not mandated but have been shown to benefit the occupants with financial savings from the efficiency of the units (Zhao et al., 2018). The low-income rent isn't based on an individual tenant's income, but rather on the 30% ceiling. An individual tenant's income is relevant only to (i) determine if they initially qualify as a low-income tenant, and (ii) determine if the developer needs to make more affordable units available if the tenant's income increases.

If a low-income tenant increases its income up to 140% of the income limit, it may still stay in the unit at the below-market rate with no other consequences to the developer. However, if its income rises to more than 140% of the limit, then the "next available unit rule" comes into play. Under this rule the developer must rent the next available unit (of comparable size or smaller) to a new low-income qualified tenant at the below-market rate. This is done because the program wants to encourage low-income tenants to increase their incomes (which may not occur if they knew a higher income could cost them their below-market rent), while at the same time still making the same number of units available to low-income households.

Once a project is built, the LIHTC property must comply with all LIHTC and project agreement terms for a 15-year compliance period. If the property falls out of compliance, investors can be subject to the recapture or loss of credits, including credits that were claimed while the project was still in compliance. For example, if non-compliance occurred in Year 14, credits in Year 1 may be subject to recapture. Following the initial compliance period, a project operates under an "extended use

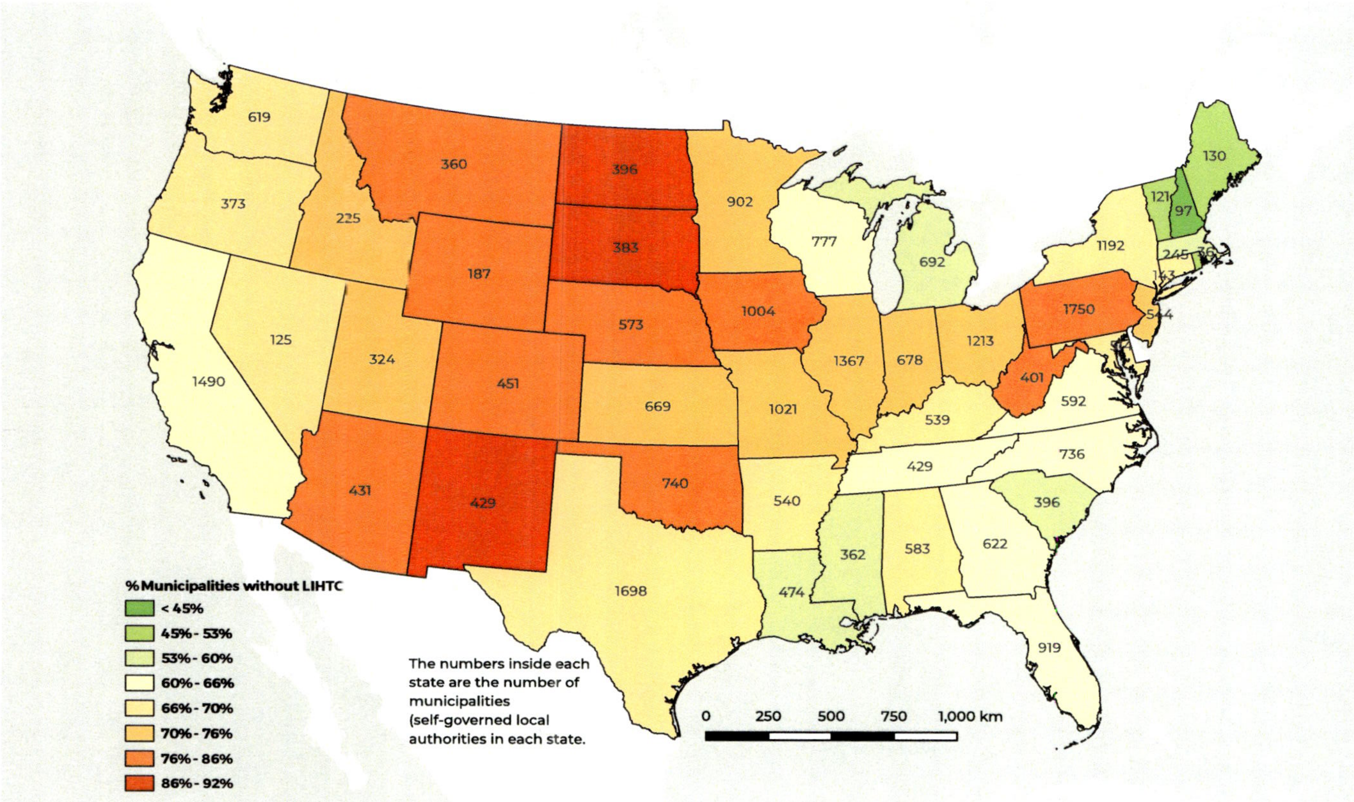


**FIGURE 3.17** LIHTC Landscape: mapping municipalities in the United States without Low-Income Housing Tax Credit (LIHTC) projects

period" (EUP) of at least 15 years (states' QAPs may require a longer EUP, e.g., California has a 55-year EUP). During this period the project must continue to provide affordable housing, but the definitions of affordable housing and compliance may differ from the definitions required during the initial 15-year period. Such definitions and other terms are negotiated and included in an EUP agreement between the state and developer.

The expiration of Low-Income Housing Tax Credit (LIHTC) poses challenges to the preservation and development of affordable housing across the country. Without the incentive provided by LIHTC, there is a risk of decreased affordable housing supply and increased financial burden on low-income individuals and families. Figure 3.18 is a map of NYC lost LIHTC units that demonstrates a concerning concentration of losses primarily in Manhattan and The Bronx. This indicates a potential impact on affordable housing availability and highlights

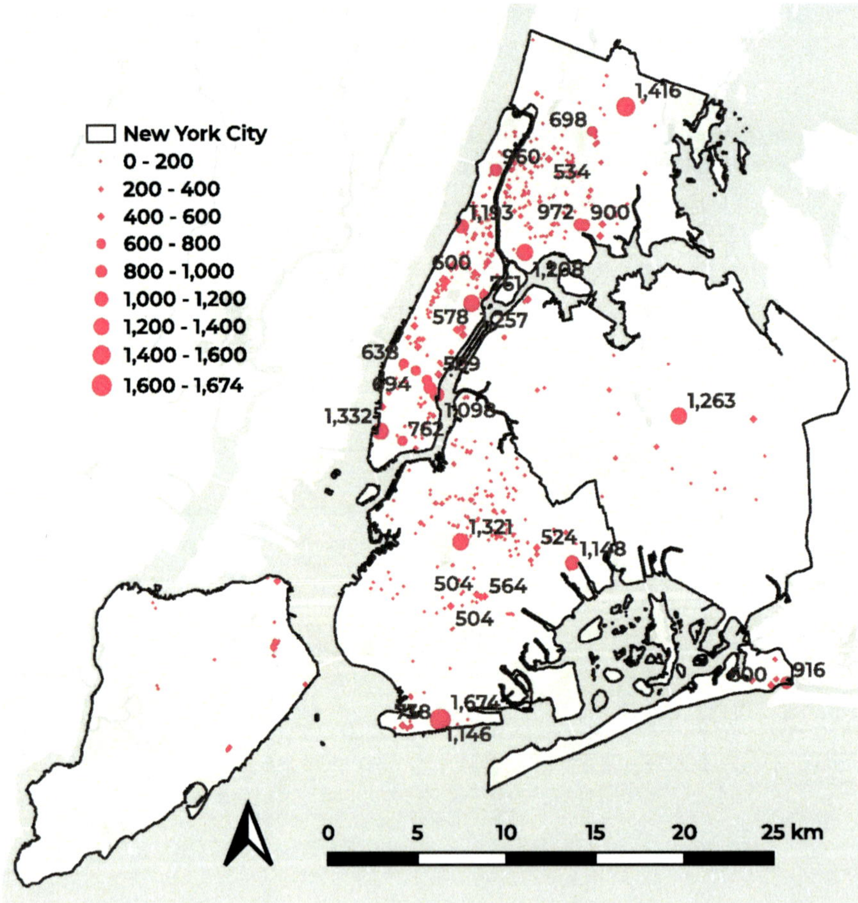


**FIGURE 3.18** Vanishing Affordable Housing: mapping Lost Low-Income Housing Tax Credit (LIHTC) units in New York City

**TABLE 3.2**
**Expiring LIHTC Properties**

| | Expired 2020 | Expires between 2020 and 2030 | Expires after 2030 | Total |
|---|---|---|---|---|
| HUD financing/insurance | 10 (11%) | 17 (18%) | 66 (71%) | 93 (100%) |
| Project-based rental assistance | 147 (31%) | 158 (34%) | 164 (35%) | 469 (100%) |
| LIHTC | 0 (0%) | 415 (28%) | 1,083 (72%) | 1,498 (100%) |
| Multiple programs | 0 (0%) | 3 (1%) | 252 (99%) | 255 (100%) |

the need for targeted efforts to address this issue in these areas. Table 3.2 (data source, HUD 2022) outlines the expiration of the LIHTC units compared to other assistance programs.

### 3.3.6 Price Restricted Housing

In large cities like New York City, several programs actively support household stability through the provision of price-restricted housing. For example, the New York City Department of Housing Preservation and Development's Housing Development Corporation offers affordable housing lotteries in new housing developments, about 5% are set aside for residents with mobility impairments and 2% for residents with visual and hearing impairments. This process is managed through an online process; the *Housing Connect web portal* allows prospective residents to create a profile and apply for a housing lottery. If selected and eligibility is confirmed, the resident can sign a rental lease or complete a purchase agreement. There is an income cap for these lotteries. Privately owned buildings have both rental and ownership opportunities. Ownership opportunities are typically in the form of a cooperative. Eligibility includes 12 months of positive rental history and meeting income requirements (for a family of 4, eligible incomes range from $0 to $220,110). The rent paid at the winning lottery buildings is determined to be affordable if it is below 33% of the individual's annual income. The program is designed for a wide range of household sizes and income levels. The income eligibility is from 0% to 30% of the federal area median income (AMI) to 165% AMI (NYC Housing Preservation and Development HPD, 2023).

The Section 32 Homeownership Program is a federal policy that allows first time homebuyers who are at or below 80% of the AMI to receive a 20% discount of home's appraised value, along with guidance to navigate the home-buying process and may include grants to cover down payments and closing costs, a one-year home warranty and lower monthly payments.

HUD's Scattered Sites Housing Programs have been in place for over five decades, serving to disperse and deconcentrate public housing in cities with dense public housing clusters. These programs create low density housing (generally under fifteen units) in middle-income neighborhoods. Scattered Site programs can be managed by city agencies and nonprofit organizations. Many cities have successfully used scattered sites programs to create permanent supportive housing. Unfortunately, there is more demand for housing than supply creating long wait times.

The Mitchell-Lama program is a unique program serving both New York State and New York City. It is intended to create affordable housing for middle class households, and is named after its sponsors, two elected officials, State Senator Mitchell and Assembly person Lama who established the program in 1955. The program has been very successful in creating stable affordable housing through cooperatives and rentals. The original program is no longer active, but it is estimated that over 100,000 apartment units were created as a result of the program. Developers were able to delist their apartments from participation in the program after a 20-year period which impacts the availability of affordable housing, see Figure 3.19 (NYCHPD, n.d.).

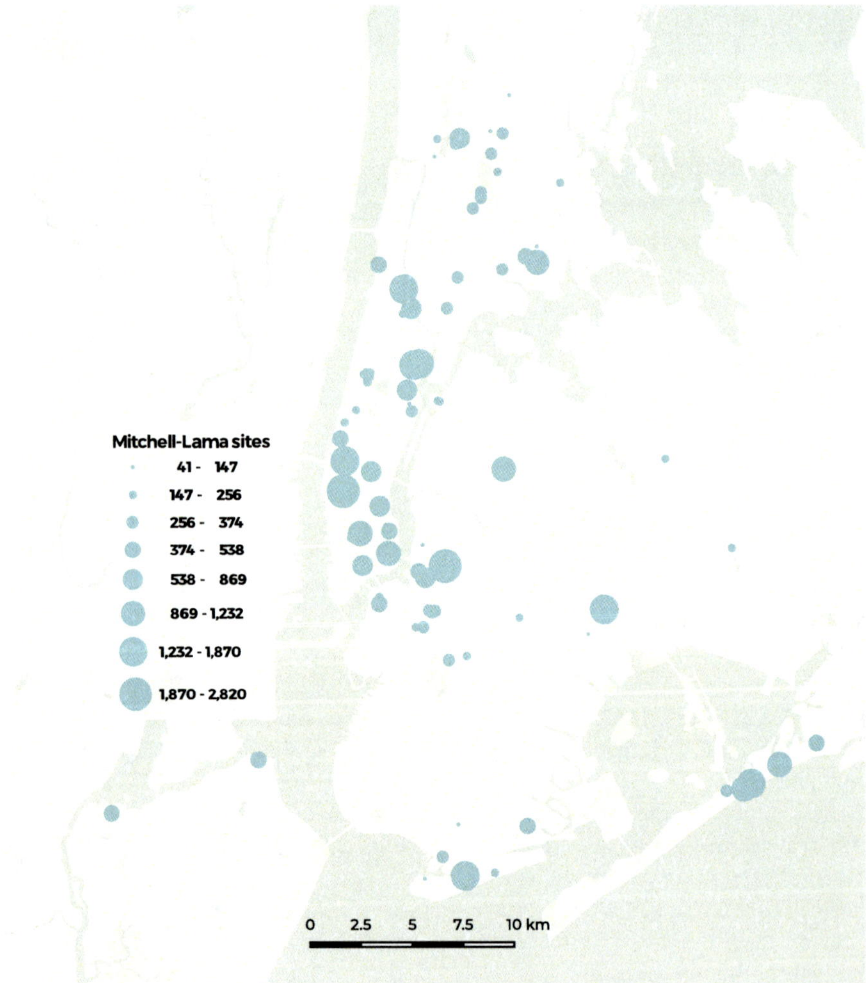


FIGURE 3.19 Mapping Mitchell-Lama development sites in New York City

## 3.4 EMERGING TRENDS IN ALTERNATIVE HOME OWNERSHIP – SHARED EQUITY HOMEOWNERSHIP

Condominiums, cooperatives, and community land trusts are forms of shared equity homeownership. These models provide mechanisms that make it easier to avoid speculative practices in real estate which can drive up housing prices and create instability. Condominiums began in the 1960s in Manhattan high-rise buildings. By sharing the ownership and responsibilities, condominiums, cooperatives, and land trusts create a community for the betterment of the entire building or complex. The boards or associations that govern these shared equity communities are responsible for looking after the interests of the whole, ensuring maintenance, management, and financial stability of the properties.

Today, new models for shared equity homeownership are emerging in the form or resale-restricted, owner-occupied housing, community land trusts (CLTs), limited equity cooperatives (LECs), and price-restrictive houses and condominiums with 30 plus years affordability covenants (Davis, 2018). These homeownership models are an alternative to single-family homes, offering diverse housing options and an opportunity for individuals and families to become homeowners while benefiting from shared amenities, reduced maintenance costs, and a cooperative living environment.

### 3.4.1 CONDOMINIUMS AND COOPERATIVES

Condominiums and cooperatives have many similarities but are different in their ownership structure and the rights and responsibilities of the residents. A unit owner of a condominium has direct ownership of their unit and holds a deed to the unit directly. Common areas are owned collectively by the unit owners through an association or cooperation. Decisions in a condominium unit are made by individual owners but must operate within the rules set by the condo associations bylaws. Alternatively, members of cooperatives, or coops, do not own their individual unit. They own shares or memberships in the coop corporation, which owns the entire building. Each resident holds a lease, which allows them to live in a specific unit. Coop residents have voting rights and can participate in its governance. Major decisions are made collectively by a co-op board of directors or general assemblies, where residents have the opportunity to voice their opinion and vote on important matters.

The structure of a condominium and cooperative may evoke high-rise buildings in New York City. Figure 3.20 (MapPLUTO, n.d.) shows the square footage of condominiums in New York City. Manhattan has a high density of large condominium buildings with very large square footage, while the outer boroughs have both these large buildings as well as many smaller buildings. While high-rise buildings are prevalent in cities like NYC, cooperative (coop) and condominium (condo) housing typologies can also be found throughout the United States in the form of semi-detached townhouse buildings within expansive complexes, often gated communities. These communities provide various amenities like community pools, fitness centers, tennis courts, or golf courses. Given that 88% of all housing structures in the United States consist of attached or detached single-family homes, opting for shared equity homeownership models such as coops and condos presents an alternative route to

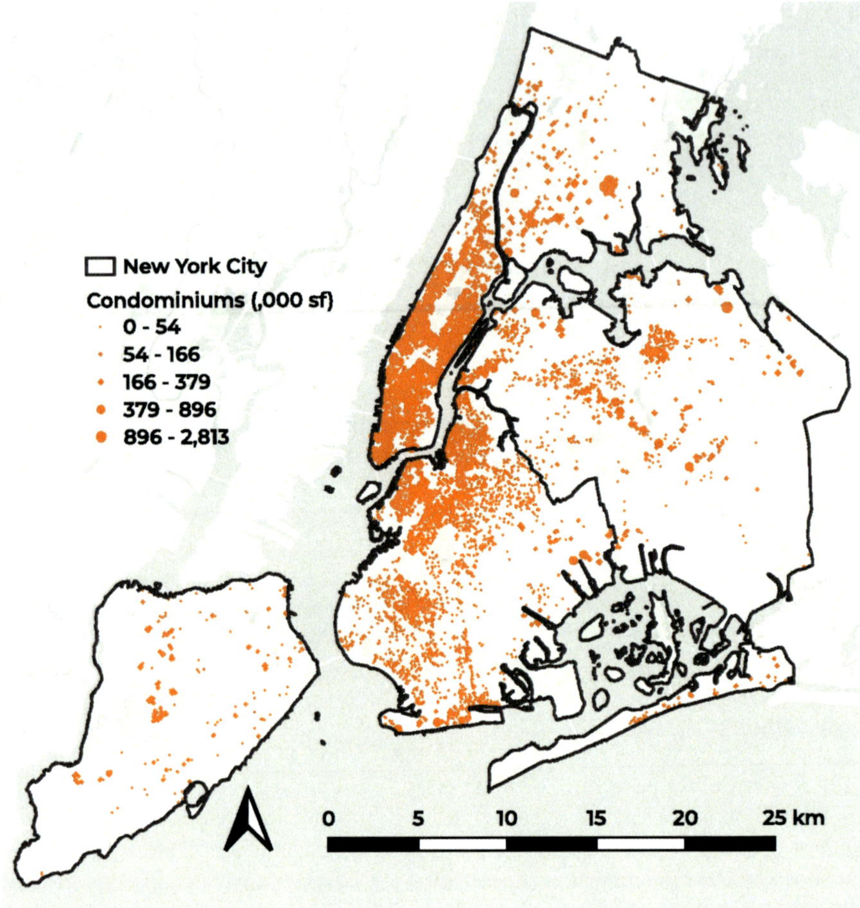


FIGURE 3.20 Mapping the concentration of condominiums in New York City

homeownership that may be more feasible for those who would otherwise find it unaffordable, see Figure 3.21 (data source, US Census 2023b).

### 3.4.2 Community Land Trusts

Community land trusts (CLT) are private entities that purchase property, usually in neighborhoods that have blight, in order to be able to lease land at set prices for the future. It is a "social invention designed to address social problems" (Meehan, 2014). CLT's ownership can be made up of community residents, non-residents, and representatives with a public interest. The idea of the CLT shifts the relation of land in the hands of a private owner to that of a community. Like cooperatives, the land is owned by the CLT and leased out to individuals. The CLT idea, however, is different from cooperatives because the trust can be made up of members that are not lease holders, but rather support the social and economic goals of the CLT.

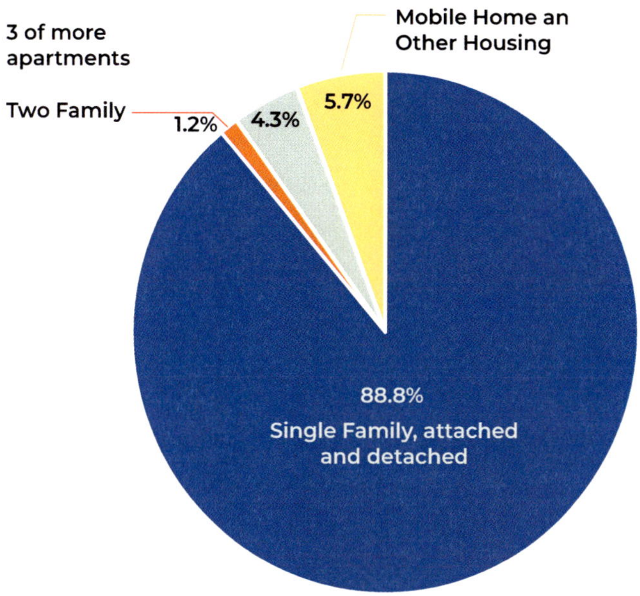


**FIGURE 3.21** Percent distribution of homeowners by structure type and number of units

As of 2023, in the United States there are about 225 CLTs. These land trusts operate at different scales but benefit the members in similar ways by creating and preserving affordable housing, stabilizing communities by preventing displacement, building wealth for low-income families, and promoting ownership and control of land. In a CLT the ownership of land does not have to be contiguous and more often the locations of the land trust are scattered throughout a community, as seen in Figure 3.22 from the Oakland CLT (OakCLT Properties, 2003). Oakland CLT has lots throughout the city of Oakland and the properties in the trust range from single-family homes, transitional housing, and commercial properties.

Community land trusts (CLTs) are a growing movement in the United States, and they are playing an important role in addressing the affordable housing crisis. CLTs can help to create and preserve affordable housing, stabilize communities by preventing displacement, build wealth for low-income families, and promote community ownership and control of land. CLTs are a promising solution to the affordable housing crisis, and they are likely to play an even greater role in the years to come.

## 3.5 USING GIS FOR STORYTELLING AND COMMUNICATION

We have emphasized the importance of spatial relationships and the need to consider housing within its geographical context. We have heavily annotated our narrative with static maps and images to communicate specific data and evidence but also to help tell a story. Static maps allow us to compare information about spatial extents (such as county boundaries or state lines) alongside the variables under consideration – for example, average home prices at a national level mask the high variability that is

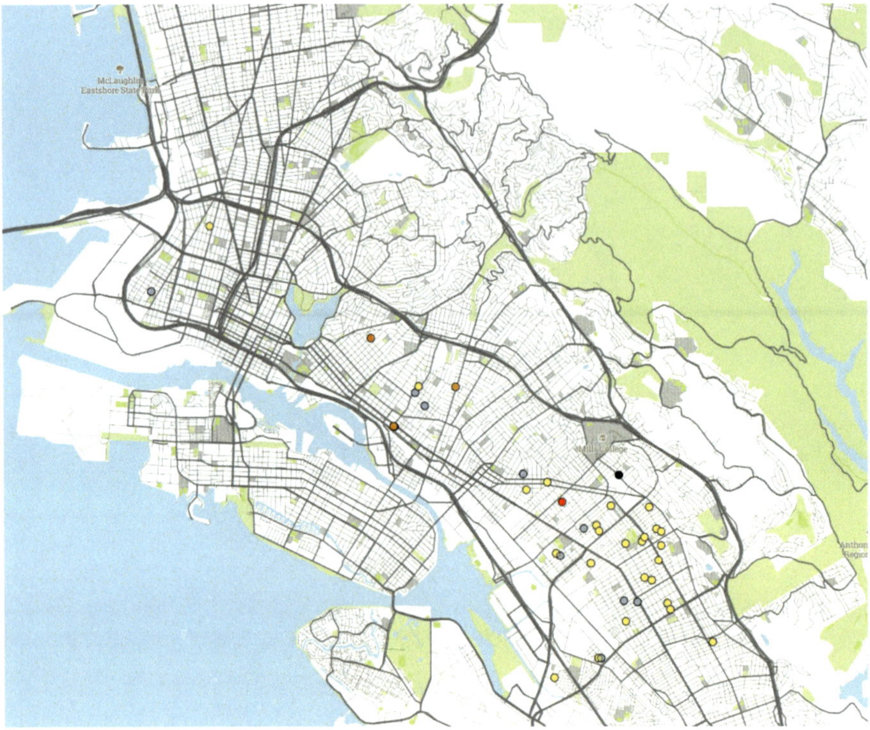

**FIGURE 3.22** The disjointed spatial layout of community land trust properties in Oakland, CA

immediately apparent when visualized at the state level. Yet, static maps alone cannot fully capture the complexity of the housing challenges we encounter. Understanding relationships between different variables anchored by the same spatial extent reveals the power of using GIS analyses. However, moving from static to dynamic representations as well as the inclusion of interactive elements are powerful ways to connect with hyper-diverse audiences. The GIS company Esri has developed two products to support impactful storytelling, but the storytelling concept can be applied even without the use of the Esri tools.

StoryMaps weave together different lines of reasoning, akin to how a statistician may likely apply multivariate analyses, but use more compelling visual narratives. For example, Steven Aviles (2022) storymap (Aviles, 2022) introduces the boom and bust phases of building construction, then visualizes housing affordability at multiple geographical scales, and concludes with a discussion of policy options. Storymaps combine emotive photography with graphics and maps that are held together by a story text of, in this case, some 2,000 words. In the Aviles example above, the section on construction history consists of three maps, one each for the westward expansion (similar to our Figure 2.2), urban sprawl (compare Figure 1.13), and first and second ring of suburbs (see our Figure 2.25). The section on affordability is supported by a table, five charts, and no less than 36 maps of eight metropolitan areas. The story ends with a discussion of zoning, ADUs and LIHTC illustrated with a graphic explaining policies aimed at densification like California's 2020 bill on subdivisions, known as *SB 1120*. Since storymaps are dynamic web pages, it is easy to enrich them with external links and embedded content.

The author of this very comprehensive storymap had the advantage of being able to work with detailed nationwide datasets compiled by their employer Esri. This allowed him to create uniform maps for the eight metro areas without having to go through the data assembly strategies that we are going to discuss in Chapter 4. A more typical storymap would tell the same story for just one study area such as *Madison's (WI) storymap explaining densification*, and it would be fairly straightforward to create if the housing researcher accesses the data that is usually held in-house in a local or regional authority. We will revisit the notion of storymaps in Chapter 5, where we expand on their ability to communicate complex analyses to local audiences. While well hidden under the shiny presentation, each storymap relies on the same data that we will use in Chapter 5 to introduce the reader to GIS analyses. The domain of housing-related data is rich and not quite self-explanatory, and for this reason, we have devoted a whole chapter on data for housing research.

## NOTES

1. Contrary to the popular image of a stand-alone house in the middle of a yard, single-family homes also include condominiums and townhouses.
2. Some states have now (2023) started to propose guidelines for fire safety in tiny homes.

## REFERENCES

ADU Handbook, 2022. *HCD Accessory Dwelling Handbook*. Sacramento, CA: California Department of Housing and Urban Development. https://www.hcd.ca.gov/sites/default/files/2022-07/ADUHandbookUpdate.pdf, last accessed 5 May 2023.

Alexander, L, 2022. "Tiny Homes: A Big Solution to American Housing Insecurity". *Harvard Law & Policy Review*, 15(2): 471–509. https://scholarship.law.tamu.edu/facscholar/1551, last accessed 21 May 2023.

Aviles, S, 2022. *How the Age of Housing Impacts Affordability*. https://storymaps.arcgis.com/stories/ac7f226a5ffd4466acbc0c7a14deab0e, last accessed 27 May 2023.

Calavita, N, and Grimes, K, 1998. "Inclusionary Housing in California: The Experience of Two Decades". *Journal of the American Planning Association*, 64(2): 150–169. https://doi.org/10.1080/01944369808975973

Calthorpe, P, and Fulton, W, 2001. *The Regional City: Planning for the End of Sprawl*. Washington, DC: Island Press.

Community Loan Fund, 2023. *New Hampshire Community Loan Fund*. https://communityloanfund.org/, last accessed 5 May 2023.

Davis, J, 2018. "More than Money: What Is Shared in Shared Equity Homeownership?" *Shelterforce*, 156(2): 30–33.

Federal Transit Administration (FTA), 2023. *Transit Oriented Development*. https://www.transit.dot.gov/TOD, last accessed 31 May 2023.

Gans, H, 1967, 2017. *The Levittowners: Ways of Life and Politics in a New Suburban Community*. New York: Columbia University Press.

Google Maps, 2023, last accessed 21 October 2023.
HUD, 2021. *Understanding SRO*. Washington, DC: US Department of Housing and Urban Development. https://files.hudexchange.info/resources/documents/Understanding-SRO.pdf, last accessed 15 May 2023.
HUD, 2022. *The Low-Income Housing Tax Credit (LIHTC)*. Washington, DC: US Department of Housing and Urban Development. https://www.huduser.gov/portal/datasets/lihtc.html, last accessed 15 May 2023.
HUD, 2023a. *Rental Assistance Demonstration: Conversion Guide for Public Housing Agencies*. Washington, DC: US Department of Housing and Urban Development. https://www.hud.gov/sites/documents/RADCONVERGUIDEPHA.PDF, last accessed 15 May 2023.
HUD, 2023b. *The Office of Manufactures Housing Programs*. Washington, DC: US Department of Housing and Urban Development. https://www.hud.gov/OMHP, last accessed 31 May 2023.
HUD User, 2023. *Low-Income Housing Tax Credit (LIHTC): Property Level Data*. https://www.huduser.gov/portal/datasets/lihtc/property.html, last accessed 5 May 2023.
LEED, 2023. *LEED v4: Reference Guide for Neighborhood Development*. Washington, DC: U.S. Green Building Council. https://www.usgbc.org/guide, last accessed 28 May 2023.
Lowenkron, H, 2021. *Creating More Accessible Inclusive Buildings*. New York: Bloomberg City Lab + Equality. https://www.bloomberg.com/news/features/2021-08-18/how-universal-design-creates-inclusive-infrastructure, last accessed 21 May 2023.
Manufactured Housing Institute (MHI), 2023. *2023 Manufactured Housing Facts: Industry Overview*. https://www.manufacturedhousing.org/wp-content/uploads/2023/06/Industry-Overview.pdf, last accessed 11 May 2023.
MapPLUTO, n.d. *New York City Department of Urban Planning and Development parcel-level database*. Online resource available at https://www.nyc.gov/site/planning/data-maps/open-data/dwn-pluto-mappluto.
McCammant, K, and Durrett, C, 2011. *Creating Cohousing: Building Sustainable Communities*, 3rd edition. Gabriola Island, BC: New Society Publishers.
Meehan, J, 2014. "Reinventing Real Estate: The Community Land Trust As a Social Invention in Affordable Housing." *Journal of Applied Social Science*, 8(2): 113–133.
Mukhopadhhyay, J, Ore, J, and Mende, K, 2019. "Assessing Housing Retrofits in Historic Districts, in Havre, Montana". *Energy Reports*, 5: 489–500, doi:10.1016/j.egyr.2019.03.008
Museum of the City of New York: Byron Collection, in World History Commons, https://worldhistorycommons.org/museum-city-new-york-byron-collection, last accessed 12 July 2023.
Nadel, S, Prindle, B, and Brooks, S, 2005. *The Energy Policy Act of 2005: Energy Efficiency Provisions and Implications for Future Policy Efforts*. Washington, DC: American Council for an Energy-Efficient Economy.
NYCDCP, n.d. *Inclusionary zoning developments in NYC. New York City Department of City Planning Inclusionary Housing Designated Areas*. Online resource available at https://data.cityofnewyork.us/City-Government/Inclusionary-Housing-Designated-Areas/w83z-2kf9.
NYCHPD, n.d. List of Mitchell-Lama addresses. New York City Department of Housing Preservation and Development. Online resource available at https://www.nyc.gov/assets/hpd/downloads/pdfs/services/MLLIST.pdf, last accessed 05/17/2023.
NYC Housing Connect, 2023. *Housing Connect Web Portal*. https://housingconnect.nyc.gov/PublicWeb/, last accessed 15 May 2023.
NYC Housing Preservation and Development, 2023. *Affordable Housing*. https://www.nyc.gov/site/hpd/services-and-information/find-affordable-housing.page, last accesses 15 May 2023.

OakCLT Properties. *OakCLT Properties*. 2023. https://oakclt.org/about/oakclt-properties/, 13 July 2023.

Parolek, D, 2020. *Missing Middle Housing: Thinking Big and Building Small to Respond to Today's Housing Crisis*. Washington, DC: New YorkIsland Press.

RAD, 2023. *Properties Participating in RAD Program*. RAD Resource Desk. https://www.radresource.net/pha_data.cfm, last accessed 5 May 2023.

Sanguinetti, A, 2015. "Diversifying Cohousing: The Retrofit Model". *Journal of Architectural and Planning Research*, 32(1): 68–90.

Scally, C, Gold, A, and Dubois, N, 2019. *The Low-Income Housing Tax Credit: How it Works and Who It Serves*. Washington DC: The Urban Institute.

Smith, R, and Bereitschaft, B, 2016. "Sustainable Urban Development? Exploring the Locational Attributes of LEED-ND Projects in the United States through a GIS Analysis of Light Intensity and Land Use". *Sustainability*, 8: 547. doi:10.3390/su8060547

SRO Housing Corporation. *SRO Housing Corporation*. 2023. https://www.srohousing.org/property-management.html, 13 July 2023

Talen, E, 2005. *New Urbanism and American Planning. The Conflict of Cultures*. Philadelphia, PA: Routledge.

US Census Bureau, 2017. *American Community Survey - Data Tables and Tools - Subject Tables*. Online Source https://www.census.gov/acs/www/data/data-tables-and-tools/subject-tables/, 28 July 2023.

U.S. Census Bureau, 2021. *2020 Census Group Quarters*. https://www.census.gov/newsroom/blogs/random-samplings/2021/03/2020-census-group-quarters.html#:~:text=Group%20quarters%20are%20defined%20as,not%20related%20to%20one%20another.

U.S. Census Bureau, 2023a. *Glossary*. https://www.census.gov/glossary/, last accessed 31 May 2023.

U.S. Census Bureau, 2023b. *Housing Vacancies and Homeownership*. https://www.census.gov/housing/hvs/index.html, last accessed 31 May 2023.

U.S. Census Bureau, 2023c. *Latest Data Tables of New Manufactured Homes*. https://www.census.gov/data/tables/time-series/econ/mhs/latest-data.html, last accessed 26 May 2023.

U.S. Census Bureau, 2023d. *U.S. Census Bureau Construction Spending - Characteristics*. n.d. Online Source https://www.census.gov/construction/chars/current.html, 28 July 2023.

U.S. Census Bureau and U.S. Department of Housing and Urban Development (HUD), Houses Sold by Type of Financing, Cash Purchase [HSTFC], retrieved from FRED, Federal Reserve Bank of St. Louis; https://fred.stlouisfed.org/series/HSTFC, 20 October 2023.

USHUD, n.d. US Department of Housing and Urban Development LIHTC property level data in MS Access and CSV format. Online resource available at https://www.huduser.gov/portal/datasets/lihtc/property.html.

Zhao, D, McCoy, A, Agee, P, Mo, Y, Reichard, G, and Paige, F, 2020. "Time Effects of Green Buildings on Energy Use for Low-Income Households: A Longitudinal Study in the United States". *Energy Policy*, 140: 111827. doi:10.1016/j.enpol.2020.111827.

# 4 Data for Housing Research

## 4.1 HOUSING DATA SOURCES

Most housing data is collected by organizations that have a financial stake in housing and need the data for the purposes of financial accountability. As such, housing data is generated (although not necessarily published) by everyone who has a financial stake in the housing market: lenders, insurances, builders, private, cooperative, governmental, or non-governmental entities. In addition, there are some data collections by foundations, think tanks, and academic institutions, although they are more often than not ad hoc; i.e., they tend to be compilations for a particular study rather than long-term repositories. The quintessential counterpart to these is the US Census Bureau, which has been collecting housing-related data for almost a century. Table 4.1 in the Appendix provides an overview of the range of suitable data sources.

### 4.1.1 US Census

At the moment, much of housing policy analysis conducted by planners focuses on analysis at the state level, comparing the impacts of government policies in different states, for instance, or at the level of level of counties. There are a little over 3,000 counties in the United States, and over 84,000 census tracts! Counties can be large or small, and often county-level analysis cannot provide the fine-grained spatial differentiation of phenomena that is necessary to understand policy or programmatic impacts.

In a very narrow sense, the responsibility of the US Census Bureau is to enumerate the population of the United States every 10 years for the purpose of apportionment of seats in the House of Representatives (ref). Given the size of the task, the Census Bureau harbors a large number of experts in the fields of demography and statistics. This in turn led to the request of many other government agencies to use these resources for the collection of a wide range of other data. As the core counting unit of the census is a household at a given address, housing is the next logical realm of data to be collected.

Across a myriad of censuses and surveys, the US Census Bureau collects literally thousands of variables, many of which are useful for housing policy research. What makes Census data quintessential GIS data, however, is the fact that each data point has a spatial reference, i.e., it refers to an area unit that is both unique and well specified. The Census Bureau is by law required to preserve confidentiality about the data collected, which means that for 72 years after each collection, data is published in aggregate form only. There are multiple ways that individual-level data can be aggregated and the result is an interesting relationship between spatial, temporal,

DOI: 10.1201/9780429243066-4

**TABLE 4.1**
**Data Source List/Summary**

| | |
|---|---|
| US Census Bureau | Information on homeownership rates, housing vacancies, and housing market characteristics |
| Zillow | Data on home values, rental prices, and other housing-related information |
| Redfin | Data on home sales, prices, and market trends. They also provide an API for accessing their data |
| Realtor.com | Real estate listings, property information, and housing market data |
| Federal Housing Finance Agency (FHFA) | Data on home prices, mortgage rates, and mortgage market conditions. They also maintain the House Price Index (HPI), which tracks changes in home prices over time |
| National Associations of Realtors (NAR) | Regular reports on existing home sales, home prices, and housing market trends in the United States |
| Bureau of Economic Analysis (BEA) | Data on housing investment, construction spending, and other economic indicators related to the housing market |
| Department of Housing and Urban Development (HUD) | Information on affordable housing programs, housing market conditions, and demographic data |
| CoreLogic | Real estate market information, including property values, mortgage data, and housing market trends |
| Local Multiple Listing Services (MLS) | Regional or local databases used by real estate agents to list and share property information |
| Home Mortgage Disclosure Act (HMDA) | Information on mortgage lending, including loan types, interest rates, and borrower demographic |
| National Association of Home Builders (NAHB) | Housing market data, including home construction statistics, building permits, and industry trends |
| S&P Case-Shiller Home Price Indices | Data on home prices in major metropolitan areas across the United States |
| Federal Reserve Economic Data (FRED) | Economic and housing-related data, including housing starts, building permits, and mortgage rates |
| Urban Institute | Datasets related to affordable housing, housing market dynamics, and housing finance |
| Mortgage Bankers Association (MBA) | Data on mortgage applications, refinancing activity, and mortgage market trends |
| Federal Housing Administration (FHA) | Data on government-insured mortgage loans, including loan volumes, delinquency rates, and borrower demographics |
| Local and regional government websites | Information on property taxes, housing permits, and neighborhood statistics |

and attribute specificity requiring end users to make some choices of how to set up their data queries: Census data can be very detailed but would then be representative only for large areas and somewhat outdated, or can be very specific to a subset of a neighborhood but only for a few common variables and again at the price of low currency, or it can be collected every month but only at the spatial resolution of counties,

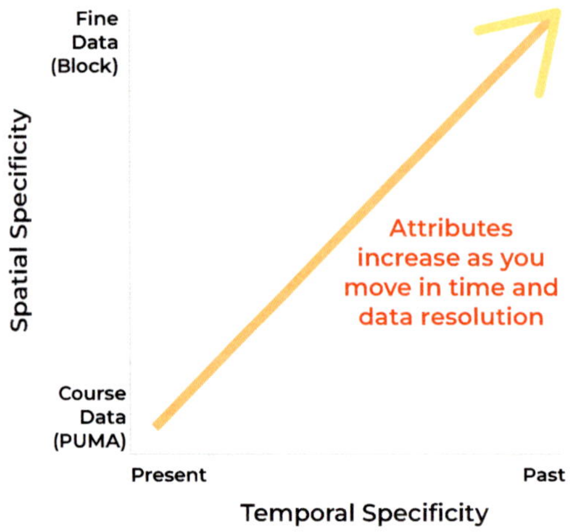


**FIGURE 4.1** Choosing between spatial, temporal or attribute specificity

i.e., some 3,000 data points for the whole country. Figure 4.1 describes this conundrum of how to navigate between the three opposing characteristics.

The temporal resolution of Census data ranges from being reported monthly (e.g., employment statistics), to yearly, 3-yearly, 5-yearly, and only once in a decade. The spatial resolution is more complicated as illustrated in Figure 4.2. The lowest level of aggregation (or in Census parlance summary level) is a census block, an area unit that on average captures some 400 people and aims to be delineated by topographic features such as a street block. Only a few very common variables are released once every 10 years at this fine spatial grain. Typically, three or four census blocks are then aggregated to establish block-groups and some variables collected over a span of 5 years are published at this level. Most, though not all, variables are available at the next higher level of spatial aggregation, the census tract. And so it goes up the ladder of Figure 4.2. All the area units along the central spine of this figure fit neatly into each other, i.e., their boundaries never intersect or cross. As more and more other government agencies asked for aggregations according to their needs, the Census Bureau also publishes data in area units such as municipal, school district, or ZIP code area boundaries, in other words – special purpose boundaries that are useful for management and governance.

In addition to the decadal Census of Populations and Households that was mentioned previously, the Census Bureau conducts a continuous American Community Survey (ACS), the results of which are published in 1-, 3- and 5-year intervals (aggregates) with gradually increasing levels of spatial specificity as the data is aggregated over longer time spans. While a census aims to be a complete enumeration of all entities of its universe (here, people or households), a survey (even one as large as the ACS) is based on a sample of the statistical population and results are estimates. Therefore, all ACS data releases are accompanied by a reference to confidence ranges. Each of the variables of these products is independent; i.e., if we have small area data for a specific time span that provides information about income and rents then we *cannot* combine these variables to deduce a causal link between these variables, that is we cannot establish the number of people in one income group category that pays a particular amount of rent. The Census Bureau

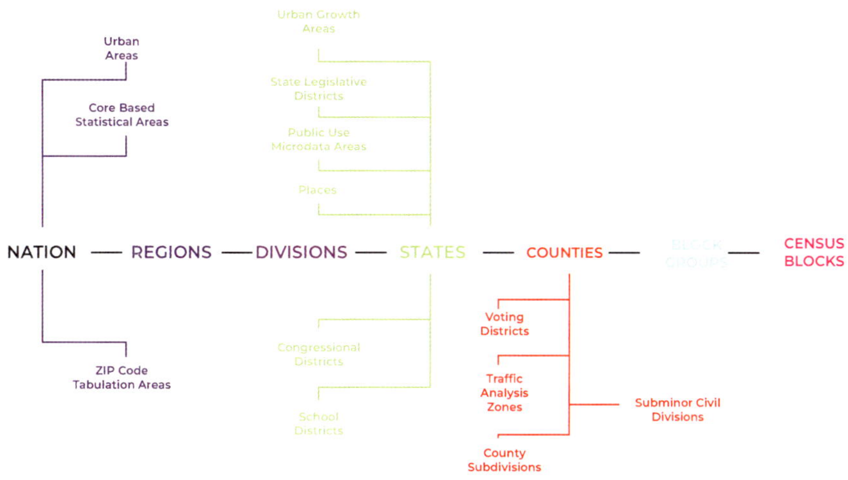


**FIGURE 4.2** Hierarchy of Census area units

does perform such calculations based on individual-level data, but such combinations of variable values are then only released at much coarser spatial resolutions (so-called Public Use Microdata Areas (PUMA) are designed in such a way that each PUMA has no less than 100,000 but often as many as 200,000 people). In addition to these very large and comprehensive products, the Census Bureau conducts many dozens of other more specialized data collections such as the American Housing Survey, Consumer Expenditure Survey, Housing Vacancy Survey, Annual Business Survey, Annual Survey of Public Employment and Payroll, Annual Survey of State and Local Government Finances, Building Permits Survey, the Census of Governments, the Economic Census, or the Survey of Construction among many others. The sheer volume of data makes the Census website somewhat difficult to navigate. Dedicated third-party websites that transform Census datasets to make them accessible for diverse audiences include *Social Explorer*, Esri's *Living Atlas,* or the *Census Reporter* discussed in the following section.

### 4.1.2 Census Reporter and Social Explorer

The US Census Bureau's web site requires a good understanding of the types of data collections, the area units, and the intricacies of attributes. Many of the datasets are unwieldy, containing hundreds of columns. This is great for expert users, who typically use application programming interfaces (APIs) to access the data they need quickly. Casual or novice users tend to get intimidated. To serve these constituents and to create access and equity, non-for-profit organizations and academic institutions have created web sites that provide the user with the results of commonly run queries and reformat the output into easily digestible spreadsheets and a number of exportable GIS formats such as KML, GeoJSON, and Geopackage. It is useful to note that these efforts have been underway for many decades and the data offerings, and the data provided have co-evolved with hardware and software advances.

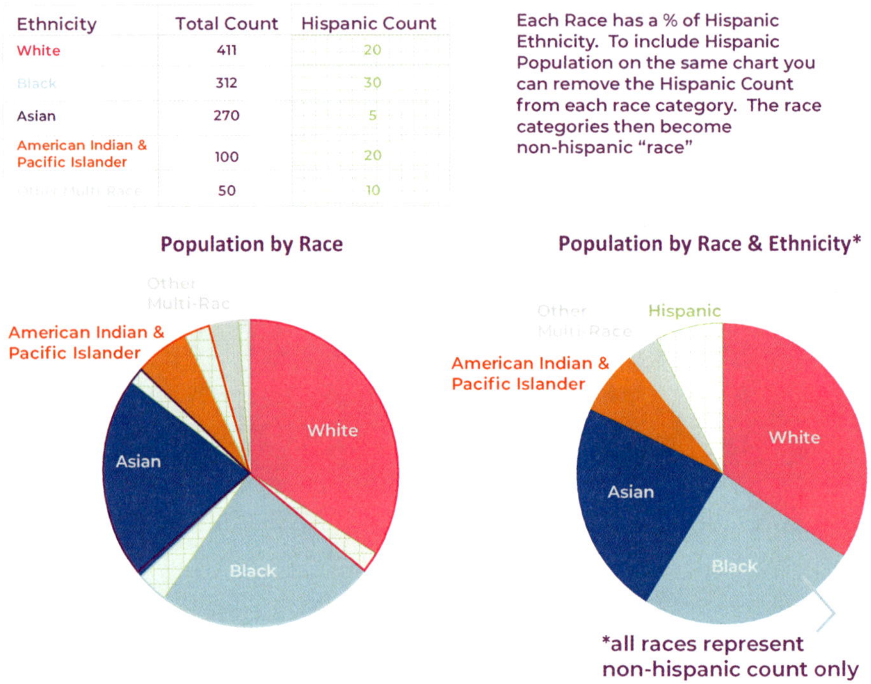


**FIGURE 4.3** Misinterpreted identities: unraveling the consequences of misaligned race and ethnicity data reporting

The *Census Reporter* is a non-profit organization that strives to make data from the American Community Survey easier to use. The Knight Foundation funded the initial build-out of the site, which is now maintained by Northwestern University's School of Journalism and hosted by Oregon State University. In addition to precompiled profiles for over 20 topics, the site provides tutorials on the Census geographies, table organization, and technical background that help site visitors to slowly transition from the *Census Reporter* to work directly with the US Census website. For each of the topic areas, the site provides not only data but means to generate graphics and maps, which web site visitors can then download or embed in their own website.

Let us consider the one topic area of interest to us: Housing! The *Census Reporter* describes it as follows.

> The American Community Survey gathers extensive data about the housing conditions of respondents, including whether they own or rent their home, how much they spend on housing, and the physical characteristics of homes. Most of the tables count the number of housing units for a given characteristic. However, a few tables estimate the number of people living in owned or rented housing units. A housing unit is anything from a house to an apartment or even a boat if a person is currently living there
>
> *(US Census 2020 FAQ).*

Every housing unit is recorded as either occupied or vacant. Some vacancies are market related, such as houses for sale or apartments for rent. Other housing units are seasonally vacant. Occupied housing units in the ACS are split into two categories: renter-occupied and owner-occupied. This distinction is known as tenure.

The appendix contains a number of lengthy tables that illustrate the sometimes overwhelming wealth of ACS data. As we will discuss in some detail in Section 4.5, the selection of data should be based on one's conceptual model and research question. For example, we may want to look at measures of neighborhood stability. If this is the case, one of the first ACS variables to look at would be geographic mobility. Conventional wisdom has it that rented housing units see more of a turnover than owned properties – with associated assumptions about housing quality or even crime. But is this true? In New York City, for instance, there are rent-stabilized neighborhoods that result in tenants staying for many decades while gentrifying neighborhoods experience significant amounts of flipping, i.e., buyers purchase the property as a real estate investment instead of a residence. The ACS provides us with a number of variables in both tenancy categories that help us to investigate the question, and the answer is of course varying from one real estate market to another – often even within a single county or city. In addition to the geographic mobility variable, which can be reverse-interpreted as what percentage of an area unit's population has been staying in place for a certain number of years, we could look at mortgage status (Table 4.3), where a low number of mortgaged properties are either a function of an old (and stable) housing stock or of flipping (which is financially more lucrative when the property is purchased with cash, thereby avoiding the interest costs). Stable residential neighborhoods are marked by stable home values, i.e., no rapid value changes when compared to those in the vicinity. The ACS housing value variables are given in Table 4.2. Housing affordability is not well captured by mere rent or purchasing costs. Both tenancy types have associated costs such as maintenance, utilities, insurance, taxes, etc. Each of these may (but don't necessarily do) add significantly

**TABLE 4.2**
**NHGIS GIS File Availability**

| | 2020 | 2012–2019 | 2011 | 2010 | 2009 | 2000 | 1990 | 1980 | 1950–1970 | 1910–1940 | 1790–1900 |
|---|---|---|---|---|---|---|---|---|---|---|---|
| Nation | X | X | X | X | X | X | | | | | |
| Region | X | X | X | X | X | X | | | | | |
| Division | X | X | X | X | X | X | | | | | |
| State | X | X | X | X | X | X | X | X | X | X | X |
| County | X | X | X | X | X | X | X | X | X | X | X |
| Census tract | X | X | X | X | * | X | X | X | X | X | |
| Block group | X | X | X | X | * | X | X | | | | |
| Block | X | | | X | | X | X | X | | | |

*Source:* https://www.nhgis.org/data-availability.

*Census tract and block group boundaries derived from the 2009 TIGER/Line files are available, but NHGIS identifies these boundaries with 2000, not 2009, because they do not completely correspond to the units used in 2009 ACS tables.

**TABLE 4.3**
**NHGIS Crosswalk Availability**

| Source Zones | Target Zones | 1990–2010 | 2000–2010 | 2010–2020 | 2020–2010 |
|---|---|---|---|---|---|
| Blocks | Blocks | X | X | X | X |
| Block group parts | Block groups | X | X | | |
| Block group parts | Census tracts | X | X | | |
| Block group parts | Counties | X | X | | |
| Block groups | Block groups | | | X | X |
| Block groups | Census tracts | | | X | X |
| Block groups | Counties | | | X | X |

*Source:* https://www.nhgis.org/data-availability.

to the overall housing costs, which the ACS captures both as a percentage of household income or by building age. The latter shifts emphasis from people to housing characteristics such as the number of bedrooms, plumbing, or heating as given in Table 4.3 in the appendix. Table 4.2 in the appendix provides an overview of the tenure variables.

The ACS records estimated selling prices for housing units under "value". It is important to remember that the values are calculated from self-reported estimates of occupied units and vacant units on the market. Housing statistics often suffer from the mixing of different data sources when it comes to such value estimates. As long as one stays with one data source (such as the ACS), data are comparable across time and geography. But ACS values should not be mixed with tax assessments, or the values calculated by licensed assessors for the purpose of securing a loan. Therefore, estimates may become less reliable in a fluctuating or falling housing market. ACS table B25081 (Mortgage Status) records the type of mortgage on owner-occupied housing units. This includes first mortgages, second mortgages, and home equity loans. A simpler classification, "with/without", is used in tables relating mortgage status to topics like real estate taxes, household income, and tenure.

Homeowners with and without mortgages have ongoing monthly costs, and the American Community Survey gathers data about these costs, which are reported in tables referring to "selected monthly owner costs". The costs are reported as either a percentage of the household income or the number of housing units in a monthly cost range such as "$1,250 to $1,499". The selected costs used for these estimates are:

- payments for mortgages, or other debts on the property
- real estate taxes
- fire, hazard, and flood insurance
- utilities (electricity, gas, and water and sewer)
- fuel (oil, coal, kerosene, wood, etc.)
- monthly condominium fees (when applicable)
- mobile home costs (when applicable)

There are two main categories for rent, contract, and gross. Contract rent is the monthly rent agreed to without adjustments for utilities or other payments. Gross rent is similar to selected monthly owner costs. It is the sum of contract rent and the average cost of the utilities (electricity, gas, and water and sewer) and fuels (oil, coal, kerosene, wood, etc.).

The ACS records the number of bedrooms for each housing unit and provides tables to relate that number to tenure and rent. Housing units with only one room are listed as having no bedrooms. The lack of complete facilities for housing units is recorded in two areas: kitchen and plumbing. A complete kitchen requires:

- a sink with a faucet
- a stove
- a refrigerator

Complete plumbing requires:

- hot and cold running water
- a bathtub or shower

If a housing unit doesn't have one of those items, it is recorded as lacking a complete kitchen or complete plumbing facilities.

Under the rubric of "Selected Conditions", the ACS describes substandard housing such as:

- incomplete plumbing or kitchens
- overcrowding
- 30% or more of the household income spent on rent or monthly owner costs

### 4.1.3 National Historical GIS

The US Census Bureau has been collecting data for over 200 years but only post 2000 data can be accessed through their website. To fill this gap, the National Science Foundation funded a long-term project called the National Historical GIS (NHGIS) which is housed at the University of Minnesota. It provides free online access to summary statistics and GIS files for US censuses and other nationwide surveys from 1790 through the present

- County and state census tables since 1790
- Census tract tables since 1910
- Tables for all original census summary levels, down to census blocks, since 1970
- Five-year periods ACS data from 2005–2009 through 2016–2020
- One-year periods ACS data from 2010 through 2019

While all of this is already impressive, NHGIS has also created a plethora of time-series tables that cover a range of basic 100%-count statistics from the 1970 to 2020 censuses as well as several popular sample-based statistics from the 1970 to 2000 long-form surveys and from ACS 5-Year Summary Files for 2008–2012 and 2015–2019. There are also tables of state and county data that go back to 1790 for Total Population and back to 1820 for Persons by Sex. Nominally integrated time series tables, which align geographic units across time by matching names and codes without regard to boundary changes, cover up to eight geographic levels ranging from the nation down to census tracts, see Table 4.2 for more information. The set of covered levels varies among tables according to which statistics are available for each level in each source year. Geographically standardized time series tables that provide estimates for a single year's geographic units by interpolating data from other years, cover 1990, 2000, 2010, and 2020 100%-count statistics for 2010 geographic units at 10 geographic levels ranging from states down to block groups.

The Census Bureau's criteria for the delineation of area units are population-based, e.g., a census tract is supposed to have appr. 4,000 residents – regardless of whether it is in New York or Wyoming. As populations grow (or shrink), the boundaries of the Census area units change, which makes it hard to compare them across years. One of the great features of NHGIS is that they provide crosswalks, i.e., definitions of area units that are consistent across the years. The extent of coverage varies among geographic units and across years. For example, census tracts covered only eight cities in 1910 and did not cover the entire United States until 1990. Table 4.3 provides more detailed coverage information. The basis for NHGIS boundaries before 2000 are 2000 boundary files. For post-2000 boundaries, it is advisable to use 2008 boundary delineations to maintain consistency across the years.

In Chapter 5, we will give examples for how to conduct analyses across years. Very few variables (such as total population) have been consistently measured across the years. Most variable definitions have been undergoing significant changes and the next section will deal with issues of categorical redefinitions. But before we get there, one final but crucial aspect of the US Census data needs to be discussed: the difference between race and ethnicity.

### 4.1.4 Race and Ethnicity in the US Census Data

The Census Bureau defines race as a person's self-identification with one or more social groups. An individual can report as White, Black or African American, Asian, American Indian and Alaska Native, Native Hawaiian and Other Pacific Islander, or some other race. As of 2000, survey respondents may report multiple races. Ethnicity determines whether a person is of Hispanic origin or not. For this reason, ethnicity is broken out in two categories, Hispanic or Latino and Not Hispanic or Latino. Hispanics may report as any race.

This has multiple confusing consequences. One is that if one adds up all racial observables, the total is larger than the total of the population because an individual may be counted multiple times in different categories. Second, ethnicity is not a racial category and should not be mingled with race counts, see Figure 4.3 which highlights this challenge. Many government statistics do not acknowledge the difference between the two variables, and we often see Hispanic being treated as a racial category. This is wrong and automatically results in faulty statistics. If one is careful, then one can use the Census tables that list the racial categories under Hispanic and

Not Hispanic and create graphics and maps that list each race with its respective ethnic subdivisions – but this is rarely done and still does not solve the issue of multi-racial self-identification. In this volume, when we use racial categories, we limit ourselves to single race declarations only. In the United States-wide context, this is an acceptable generalization, even if it ignores the approximately 10% of the population who in 2020 declared themselves multi-racial. For detailed studies, researchers have to decide whether it is acceptable to follow this approach or whether the inclusion of multi-racial counts paints a better picture of the specific situation.

## 4.2 FROM MEASUREMENT TO INDICATORS

Primary data collection starts with measurements, using manual or automated counts or by conducting surveys and trusting that people answer honestly. Each measure has a unit of measurement and an expected range of values (for example, the measurement of the number of people in a specific place may increase or decrease but will never include values below zero). The sum of all observations of a measure is a variable. Most housing datasets combine a multitude of variables into tables. In the case of geospatial datasets, one or more of the variables are a spatial reference that associates a record with a specific location.

### 4.2.1 Locational References

Locational references may come in many shapes and forms. They may be *x*, *y* or latitude/longitude coordinates, addresses, or pointers to well-defined areas such as ZIP code areas, census area units, school districts, etc. The notion of pointers suggests a division of labor, where the details of the locational reference (e.g., the coordinates that make up the boundaries of an area) are stored in one file and the actual measures (house prices, income, etc.) are stored in another file. The pointer then acts as the unique common link between the observation and the location of the observation. This method, known as the geo-relational principle, is quite common with geospatial data as it allows linking multiple datasets to the same location rather than having to store the geospatial details in every dataset (Albrecht, 2007).

Locational references are usually strings – even if the strings consist of a sequence of numbers, which confuses not just users but also many software packages reading geospatial data. ZIP codes are a widely known representative of such numeral strings, where the position of a digit has a hierarchical meaning. For example, ZIP code areas in the US Northeast start with a zero (emphasizing the fact that these are strings rather than numbers), whereas ZIP code areas on the West Coast start with the digit 9. The American National Standards Institute (ANSI) is responsible for maintaining Federal Information Processing Series (FIPS) codes and Geographic Names Information System (GNIS) codes. A wide audience uses FIPS codes and GNIS codes across many private and public datasets to uniquely identify geographic features.

This becomes a lot more important when dealing with the locational references in US Census data, where the GeoID is a fairly long string which is built up from left to right following the schema laid out in Table 4.4. They uniquely identify all administrative/legal and statistical geographic areas for which the Census Bureau tabulates data. From Alaska, the largest state, to the smallest census block in New

**TABLE 4.4**
**The Structure of a US Census GeoID**

| Area Type | GEOID Structure | Number of Digits | Example GEOID |
|---|---|---|---|
| State | STATE | 2 | 48 |
| County | STATE + COUNTY | 2 + 3 = 5 | 48,201 |
| County subdivision | STATE + COUNTY + COUSUB | 2 + 3 + 5 = 10 | 4,820,192,975 |
| Places | STATE + PLACE | 2 + 5 = 7 | 4,835,000 |
| Census tract | STATE + COUNTY + TRACT | 2 + 3 + 6 = 11 | 48,201,223,100 |
| Block group | STATE + COUNTY + TRACT + BLOCK GROUP | 2 + 3 + 6 + 1 = 12 | 482,012,231,001 |
| Block | STATE + COUNTY + TRACT + BLOCK | 2 + 3 + 6 + 4 = 15 | 482,012,231,001,050 |

York City, every geographic area has a unique GeoID. Some of the most common administrative/legal and statistical geographic entities with unique GEOIDs include states, counties, congressional districts, core based statistical areas (metropolitan and micropolitan areas), census tracts, block groups and census blocks.

The US Census Bureau uses FIPS codes which are assigned alphabetically by geographic name for states, counties, core based statistical areas, places, county subdivisions, consolidated cities and all types of American Indian, Alaska Native, and Native Hawaiian (AIANNH) areas. Lists of geographic FIPS codes in census products can be found on the ANSI/FIPS Codes page. FIPS codes for smaller geographic entities are usually unique within larger geographic entities. For example, FIPS state codes are unique within the nation and FIPS county codes are unique within a state. Since counties nest within states, a full county FIPS code identifies both the state and the nesting county. For example, there are 49 counties in the 50 states ending in the digits "001". To make these county FIPS codes unique, the state FIPS codes are added to the front of each county (01001, 02001, 04001, etc.), where the first two digits refer to the state the county is in and the last three digits refer specifically to the county.

The US Census Bureau creates and maintains geographic codes for many statistical geographic areas that are not covered by FIPS codes. These geographic areas include census divisions, census regions, census tracts, block groups, census blocks, and urban areas. The full GEOID for many levels of geography combines both the FIPS codes and Census Bureau codes. For example, census tracts, block groups, and census blocks nest within state and county; therefore, the GEOIDs for each of these geographic areas contain both the state and county FIPS codes, in which they nest. Figure 4.4 illustrates the hierarchical relationship of different geographic areas with one another. Whereas Table 4.4 shows the GEOID structure in TIGER/Line Shapefiles[1] for some of the most common legal and statistical geographies, as well as example GEOIDs for different geographic areas.

### 4.2.2 Derived Housing Variables

A conceptual model is a representation of a system. Conceptual models are often abstractions of things in the real world, whether physical or social. They consist of concepts

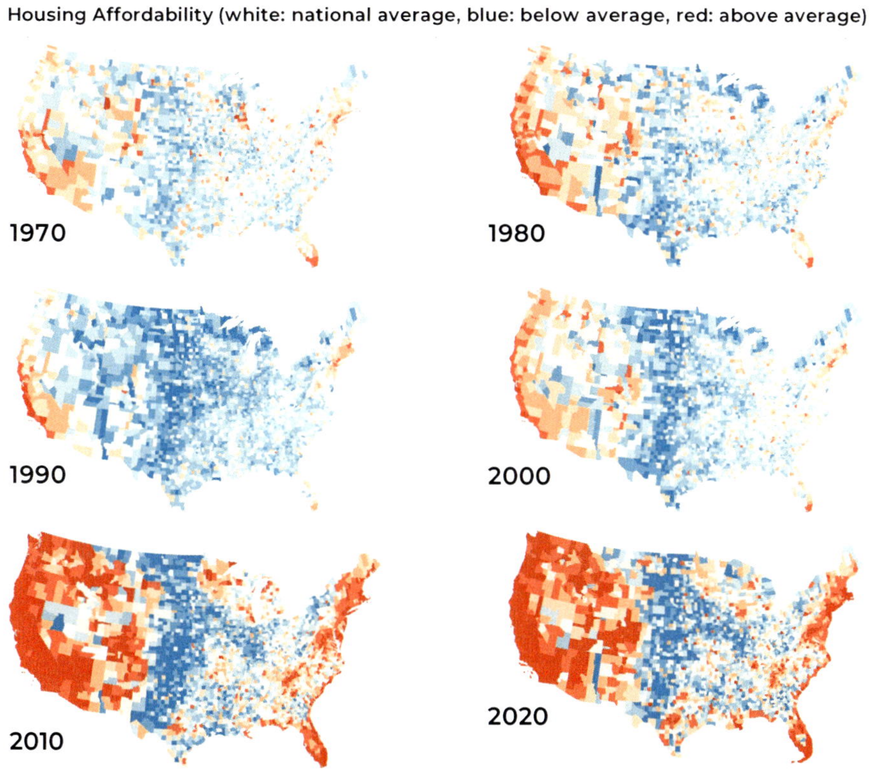


**FIGURE 4.4** Fading affordability: examining housing affordability in the US from 1970 to 2020

used to help people understand the subject the model represents. They are formed after a conceptualization or generalization process. Generalizations posit the existence of a domain or set of elements, as well as one or more common characteristics shared by those elements thus creating a conceptual model. As we acquire and explore data, we will find quite often that in spite of our utmost endeavors, it does not represent our conceptual model. A conceptual model is a representation of a system. Conceptual models are often abstractions of things in the real world, whether physical or social. They consist of concepts used to help people understand the subject the model represents. They are formed after a conceptualization or generalization process. Generalizations posit the existence of a domain or set of elements, as well as one or more common characteristics shared by those elements thus creating a conceptual model.

Typically, the conceptual model of the phenomenon we are investigating requires data that does not exist. If it exists, it may not be aligned elegantly with the conceptual model and therefore require additional manipulation or data wrangling, in industry parlance. The discussion of race and ethnicity in Section 4.1.4 is a good example of the need to transform or map variables from one census year to another. As racial categories were added, we cannot easily compare 1950s or 1970s data with their supposed equivalent from 2000 or later. We may then have to resort to inventing our own variables such as "Non-White" to appropriately represent minority populations.

In other instances, we might want to invent our own indicator for measures that do not exist in the original datasets. Housing affordability is an example of such a derived variable. There are many possible ways to define housing affordability and there are many organizations who have established their own measures. A fairly straightforward measure of housing affordability compares the median rent in an area with the median income in the same area (HUD PD&R Edge, 2017). However, even this simple approach assumes that the majority of people in that area are renters rather than homeowners – which is not the case for the majority of locales in the United States. In that case, we would have to use the current house value, compare it to the area median income, and then weigh this by a measure of how many owners have paid off their mortgage and over how many years the house price should be annualized. Technically, this is all possible, but it illustrates that simple measures can snowball quickly into complicated intractable ones if we want to paint a fair picture across the nation. It is hence the responsibility of the housing policy researcher to be very specific in the definition of their terms and the universe within which they are applicable.

One of the great advantages of working with US Census data is their consistency across the nation. Things get very complicated when we are trying to compare state- or even city-level programs, which in turn tend to have limited life spans, i.e., they are expiring and sometimes replacing each other. A good example of that is the loss of rent-regulated apartments in New York City, which we describe in the following section.

## 4.3 CHANGE OVER TIME

Although we have espoused the significance and value of understanding the spatial components of housing data, a full understanding of the phenomenon can only be gained if we conceptualize housing and neighborhood change as a spatially differentiated *process*. This requires at a minimum two timestamps for each location and ideally a lot more to capture, for example, the differential aspects of demographic, climate, economic, or policy changes across the country. Everything we observed so far in this chapter still applies but is now compounded by trying to (i) find and (ii) align data across the years. Census tract-level data is exhaustively available only since 1990 and it is hard to imagine these days how little data was collected during the last century overall, and how little of that has been properly archived and curated to be accessible today. Although we are now able to access scanned copies of the NY Times over the last hundred years, many local newspapers have ceased to exist taking their archives (of house prices, for instance) with them. The best source for historical data (with history being as recent as the 1990s) is therefore again the National Historical GIS (NHGIS). As before with the race categories, we need to be conscious of the changing definitions of the variables recorded. We will illustrate this using the example of rent/income changes over time in the following section.

### 4.3.1 Rent/Income Changes

Among the few variables that have been "consistently" collected over many decades are housing rents and income. It therefore stands to reason that we should be able to study whether housing has become more or less affordable over the years, how different the picture is in different parts of the country, and whether there are any correlations with potentially explanatory variables. Figure 4.5 provides the answer to those questions. In Section 4.9 we outline the conceptual and practical steps it takes to arrive at this figure. The six maps show clear instances of spatial autocorrelation, where likewise values are near each other rather than being randomly distributed within the study area. From a spatial analysis perspective, it would then be interesting to determine the temporal correlation and where socio-economic developments become seeds for the spatial spread of housing affordability (or the lack of) in later years. But this goes beyond the scope of this chapter.

One of the difficulties in working with Census data (despite the US Census being the most consistent and well-documented source of housing data) is that the definitions of variables change over the years. In different Census years, income is accounted for either on a per household or per family basis – and sometimes both. If nothing else is specified, then income is salaried income, excluding transfer payments as well as income from interest, stock options, etc. Tabulations for income have multiplied since the 1970s, when just about the only figure was the median income per area unit. Since then, a variety of other income-related variables have become available, e.g., social security income, aggregate income, the number of people in a particular income bracket, and so on. Similarly, rent started out as just the median rent per area unit but in later years is provided as the number of households in a particular rent bracket (which given the inflationary nature of the subject is changing from decade to decade).

## 4.4 THE AMERICAN HOUSING SURVEY

The American Housing Survey (AHS) is sponsored by the US Department of Housing and Urban Development (HUD) and conducted by the US Census Bureau. The survey provides information on the size, composition, and quality of the housing across the nation and in major metropolitan areas and measures changes in the housing stock as it ages. The AHS is a longitudinal housing unit survey conducted biennially since 1989 in odd-numbered years. While national data are always collected, typically no more than 30 metropolitan areas are sampled in one survey year. The survey includes questions about:

- the physical condition of homes and neighborhoods,
- the costs of financing and maintaining homes, and
- the characteristics of people who live in these homes.

Planners, policy makers, and community stakeholders use the results of the AHS to assess the housing needs of communities and the country. These statistics inform decisions that affect the housing opportunities for people of all income levels, ages, and racial and ethnic groups. Since the United States changes rapidly, policymakers in government and private organizations need current housing information to make decisions about programs that will affect people of all income levels, ages, and racial and ethnic groups.

HUD uses the AHS to create a biennial Worst Case Needs report to Congress, improve the efficiency and effectiveness of housing programs, and design programs appropriate for different target groups, such as low-income families, first-time home buyers, and the elderly. HUD also uses the data to allocate funds to resolve housing problems, determine qualifications for low-income housing assistance programs, and plan community development (e.g., roads and schools). Academic researchers and

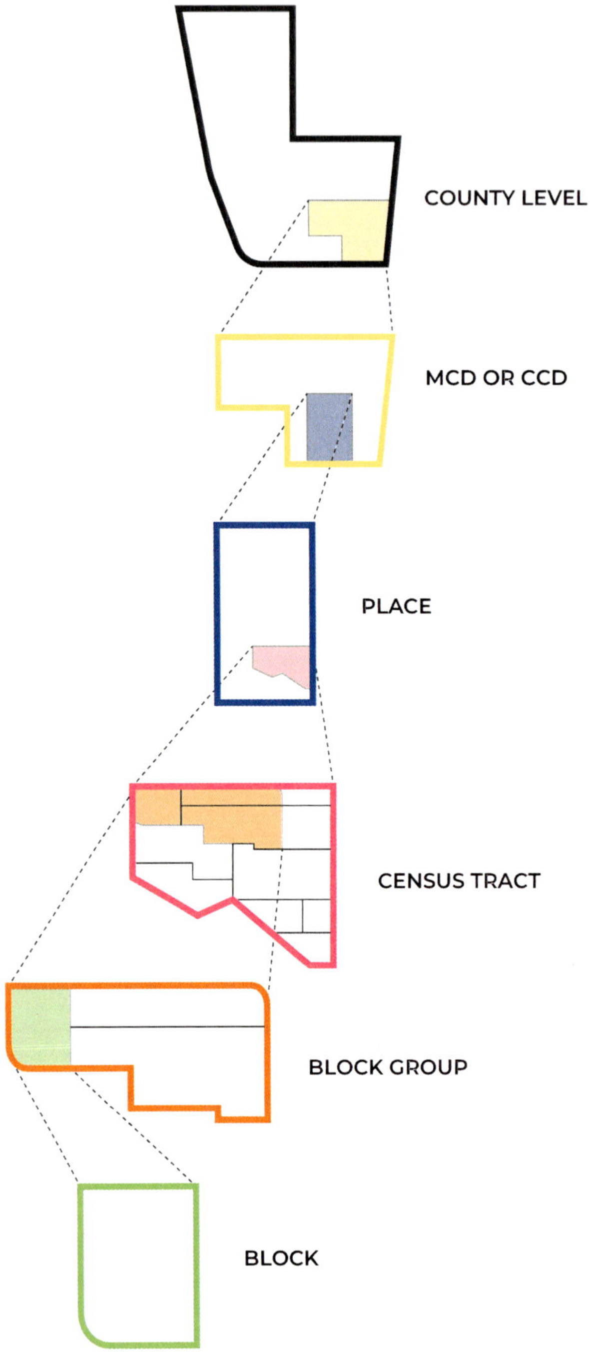


**FIGURE 4.5** Unraveling the Census Geography: exploring the interconnectedness of the US Census Bureau's geographic entities and their nested relationships

private organizations also use AHS data to analyze trends in the housing market in efforts of specific interest and concern to their respective communities.

Congress requires the Department of Housing and Urban Development to collect this information under the Housing and Urban-Rural Recovery Act of 1983 (Title 12 of the U.S.C., Section 1701z-1, 1701z-2(g), and 1701z-10a).

Beginning with the 2011 AHS, the survey instrument consists of a permanent core questionnaire plus topical supplements that will rotate in and out of the questionnaire on a yet to be determined schedule. The AHS provides current information on a wide range of "core" housing subjects, including but not limited to the following:

- size and composition of the nation's housing inventory
- vacancies,
- owners and renters,
- physical conditions of housing units,
- equipment breakdowns,
- characteristics of occupants,
- housing and neighborhood quality,
- mortgages and other housing costs,
- fuel usage,
- home improvements,
- persons eligible for and beneficiaries of assisted housing,
- characteristics of recent movers, and
- home values.

In addition to the "core" data, the AHS collected "topical" or supplemental data using a series of modules that will rotate in and out of future surveys. The 2019 topics included:

- home accessibility,
- food security, and
- post-secondary education.

The 2021 AHS includes a mortgage module redesign and the following topical contents:

- Wildfire Risk
- Household Pets
- Secondhand Smoke
- Housing Search
- Intent to Move
- Delinquent Payments and Notices

The 2015 American Housing Survey underwent a major redesign – a new sample was redrawn for the first time since 1985 and new households were asked to participate in the survey, the questionnaire was redesigned, variables were dropped, added, or modified, recodes and imputation methods were streamlined, and the weighting methodology changed. As a result, tables were redesigned, and some estimates became incomparable with previous years.

## 4.5 ESTABLISHING A GIS DATABASE FOR HOUSING PLANNING RESEARCH

GIS is commonly associated with visualization or more specifically beautifully rendered maps. What is often underappreciated is the fact that GIS relies on large and often complicated databases that reflect the complexity of geographic contexts. Whereas traditional housing research uses one dataset or the other and then represents them in the form of some business graphics, the "I" in GIS is about the (spatial) relationships between different data. The data often comes from different providers, has originally been compiled for different purposes, and in addition to the recoding covered in Section 4.2, now needs to be related to each other. Regardless of whether we are looking at metropolitan or national datasets, the resulting databases often go beyond what can be easily handled on personal computers. In any case, dedicated databases have to be created that should reflect the housing researcher's conceptual model. The common procedure to accomplish this is to build a database schema that captures all the characteristics needed – but no more. Building the final database is as much about removing unwanted variables as it is combining those we seek.

It is beyond the scope of this volume to discuss the foundations of relational database management; suffice it to state here that all the aspects of our research question need to be represented in a collection of tables that are unambiguously linked to each other. Larger organizations will do this in the form of a commercial or open-source database management system like Oracle or Postgres. But every housing researcher is encouraged to mirror the process even in smaller projects by organizing their data in personal database structures such as a geopackage or SpatiaLite.[2]

The first step in organizing one's data is to develop a conceptual model of one's research question. The most common representation of such a conceptual model is a mind map; a listing of all the important aspects of the research question and the relationships between them. If our topic, for instance, is housing insecurity, then we would want to include types of housing insecurity (overcrowding, unsafe housing conditions, eviction, and housing discrimination), factors (poverty, unemployment, rising housing costs, and lack of affordable housing), consequences (foreclosures, evictions, homelessness, housing displacement, poor health outcomes, and economic hardship), demographic groups disproportionately affected by housing insecurity (e.g., low-income households, people of color, and individuals with disabilities), and potential solutions such as affordable housing initiatives, tenant protections, and homelessness prevention programs, see Figure 4.6 for a graphic representation of this type of mind map. For some of these, we would have to determine what variables could serve as indicators, e.g., the number of times a household moves within a year, housing costs as a percentage of income, utility disconnections, or the physical state of the housing infrastructure.[3] Some of these factors are compound variables as in housing costs, which include rent/mortgage, utilities, and property insurance. For each of these factors, the housing researcher then needs to determine the unit of measurement, the spatial and temporal scale (per month or per year, per household or per county), and the likely range of observed values (for quality control purposes, see also the next section).

We recommend that the housing researcher develops this kind of a conceptual model *before* searching for the data to populate their database. There are multiple reasons for that. One is to focus one's mind on the essentials. The resulting database should contain only what we need rather than be the results of an indiscriminate data hunting and

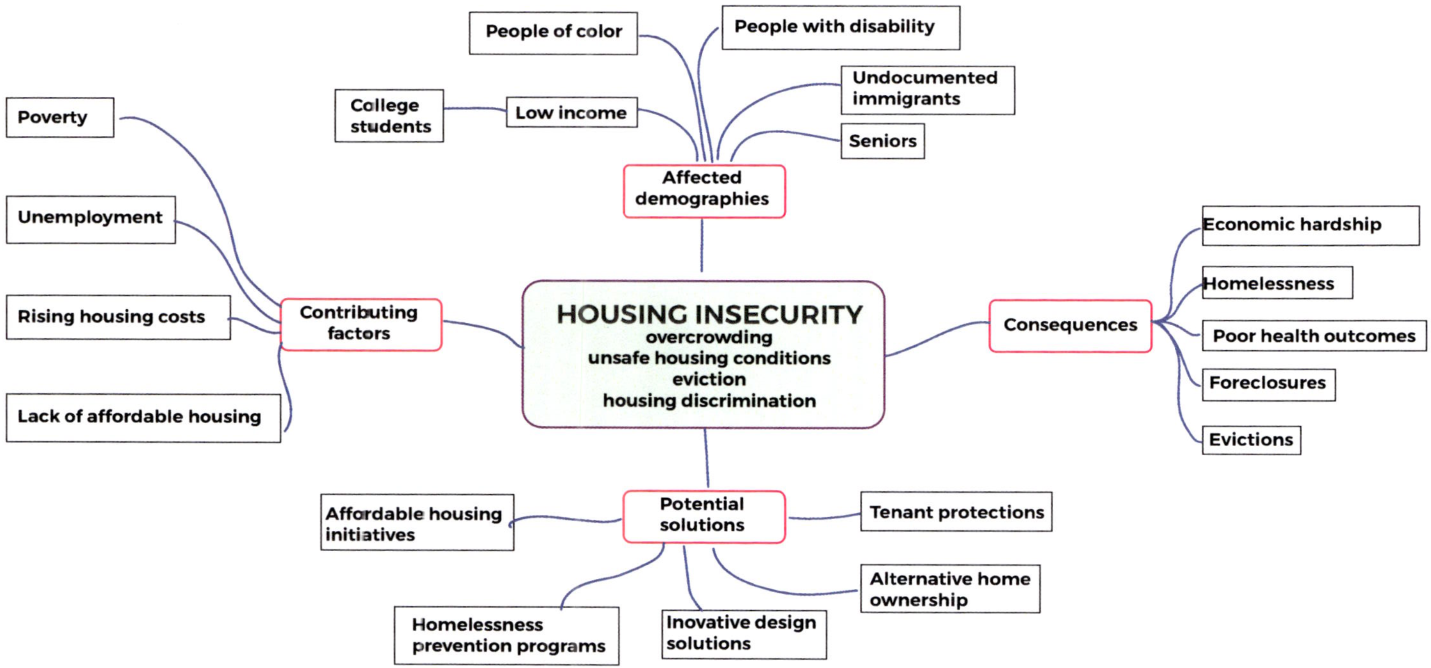


**FIGURE 4.6** Mapping housing insecurity: a comprehensive mind map unveiling the causes, effects, affected individuals, and potential solutions to address the crisis

gathering endeavor. Another very important reason is that we do not know what will be useful to us if we have not gone through the exercise of developing the conceptual model. The more rigorous our database schema is, the better equipped we are to know what data we need (to look for) and how to substitute or wrangle the data we are actually getting hold of to satisfy our needs. Finally, the difference between the idealized conceptual model and the actually filled database tells us something about how good the basis for our analysis is. Without the prior development of a conceptual model, we would not be able to judge the quality of the data we are actually dealing with.

A database schema is the translation of the mind map into an empty database structure. Each of the factors becomes a table for which we have to define what variables it consists of and what datatype is to be used for each variable. Sometimes, this is straightforward as in the setup of household income. Things get a little more complicated when we look at something like the state of the physical infrastructure (doors, windows, walls, roofs, etc.); do we want this measured on a Likert scale and if yes, at what level of aggregation (housing unit, building, census tract)? This is also the time to decide about the spatial reference: do we want the records in our tables to link to an address, an area unit, and an x/y coordinate? The database schema is the well-specified but empty shell of our database. It is defined to exactly fit our needs (which we assessed in the form of our conceptual model). Once our database schema is set up, we are ready to fill the database with data. Sometimes, this is as easy as a one-to-one import of a table into a matching (empty) table in our database. Usually, however, we will select a subset of external tables and have to transform their contents to match the specifications of our database schema. See Figure 4.7 for an example of Housing Insecurity Database Schema.

## 4.6 DATA QUALITY

As alluded to above, one of the advantages of developing a conceptual model and then designing the database schema accordingly is that any discrepancy between the idealized schema and our adaptations of that schema to match existing data is an indication for how well the data we are working with is suited to truly answer our original research question. Discrepancies between the two are captured by what is known as metadata (data about data). As we seek to fill our own database with data and search the Internet for possible data sources, the metadata tells us how close the external data matches our internal needs. If we cannot find formal descriptions of data, then this sends a warning sign that we might want to be very careful using the data we found.

Official (FGDC- or ISO-conform) metadata consists of many different dimensions of data quality: completeness, spatial, temporal, thematic accuracy, and precision, as well as consistency. In addition, a good metadata documentation will tell us by whom and how the data was generated, for what purpose, how long it is valid, and who is the custodian (from whom we might learn more about it).

Again, the US Census Bureau is the standard against which all other data sources can be measured by. For a novice user of Census data, the exhaustive description of data quality that is directly embedded into the data rather than in a separate metadata document can be stressful. Every ACS variable is accompanied by its respective Margin of Error (MoE) at the 90% confidence level. This implies a 10% chance of incorrect inference for all estimates, see Table 4.5.

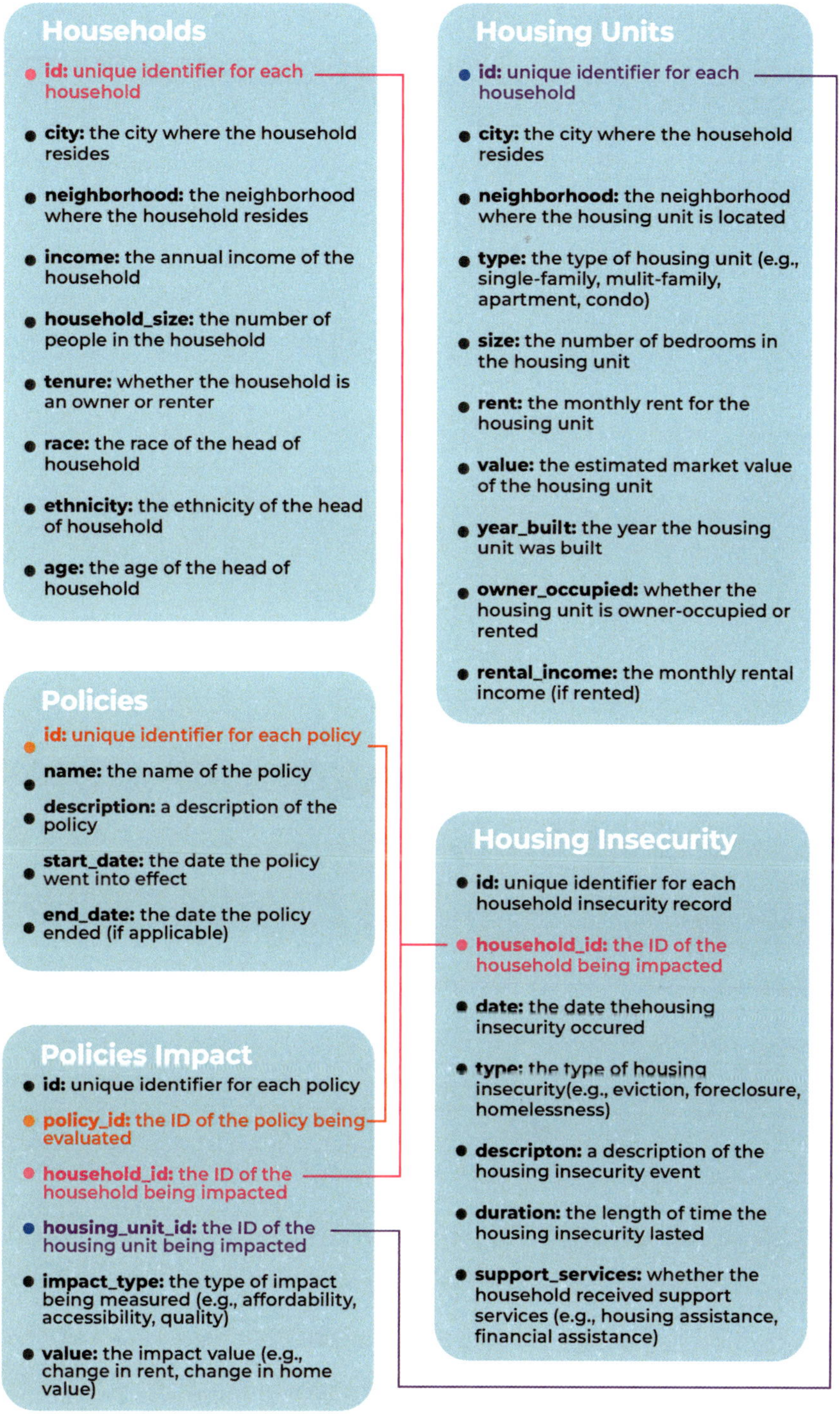


**FIGURE 4.7** A housing insecurity database schema based on the mind map of Figure 4.6

**TABLE 4.5**
**Sample US Census Table With the Margin of Error Information**

| Tract | Total | Utility Gas | Bottled or LP Gas | Electricity | Fuel Oil | Coal | Wood | Solar | Other Fuel | No Fuel Used |
|---|---|---|---|---|---|---|---|---|---|---|
| 1.01 | | | | | | | | | | |
| Estimate | 2,226 | 1,707 | 0 | 490 | 17 | 0 | 0 | 0 | 0 | 12 |
| MoE | ±302 | ±334 | ±9 | ±173 | ±27 | ±9 | ±9 | ±9 | ±9 | ±18 |
| 1.02 | | | | | | | | | | |
| Estimate | 1,591 | 847 | 0 | 699 | 0 | 0 | 0 | 0 | 13 | 32 |
| MoE | ±151 | ±133 | ±9 | ±126 | ±9 | ±9 | ±9 | ±9 | ±19 | ±32 |
| 2.01 | | | | | | | | | | |
| Estimate | 1,747 | 1,106 | 9 | 607 | 25 | 0 | 0 | 0 | 0 | 0 |
| MoE | ±184 | ±205 | ±14 | ±178 | ±35 | ±13 | ±13 | ±13 | ±13 | ±13 |
| 2.02 | | | | | | | | | | |
| Estimate | 2,415 | 1,764 | 0 | 520 | 0 | 11 | 120 | 0 | 0 | 0 |
| MoE | ±266 | ±242 | ±13 | ±175 | ±13 | ±19 | ±188 | ±13 | ±13 | ±13 |
| 3.01 | | | | | | | | | | |
| Estimate | 622 | 507 | 0 | 61 | 9 | 0 | 6 | 0 | 18 | 21 |
| MoE | ±44 | ±52 | ±9 | ±33 | ±9 | ±9 | ±8 | ±9 | ±14 | ±13 |

Most other data sources will not have such intricate quality information down to the individual record level. But at a minimum, there should be a separate metadata document (preferably following an established standard such as *Dublin Core* or *FGDC*) and a data dictionary.[4] The lack of such documentation suggests poor data quality in the first place and hence limited reliability for our data analysis down the road.

Most other data sources will not have such intricate quality information down to the individual record level. But at a minimum, there should be a separate metadata document (preferably following an established standard such as *Dublin Core* or *FGDC*) and a data dictionary.[5] The lack of such documentation suggests poor data quality in the first place and hence limited reliability for our data analysis down the road.

### 4.6.1 Data-Poor Environments

As soon as we move beyond federal data collection efforts, we will find that housing data is getting sparse. Few states and even fewer municipalities or non-profit organizations have the resources to collect housing-related data. Companies (especially utilities) are not prone to share their data and the data collected by academic organizations tends to be limited in spatial and temporal scope. The result is a patchwork of data that is impossible to generalize. In Section 4.8, we will identify a few non-conventional data sources but in the meantime, the onus is on the individual researcher to peruse data portals such as the *National Neighborhood Data Archive* at the University of Michigan, the UC Berkeley's *Urban Displacement Project*, Esri's *ArcGIS Data Hub*, the U.S. *city open data census*, or general purpose repositories such as *Awesome Public Datasets* or *Kaggle*.

## 4.7 SCALE ISSUES

We have emphasized that GIS-based housing research relies heavily on combining different datasets. The moment we do this, however, there is a good chance that the datasets have been compiled at different scales/resolutions, e.g., counties vs. metropolitan areas, or ZIP code areas vs. census tracts. One of the main functions of GIS is to overlay and disaggregate the respective datasets to create area units with aligned boundaries. We will discuss those techniques in detail in the following chapter; however, regardless of how we proceed, there are a few methodological issues that housing researchers need to be aware of, the most prominent among them being the modifiable area unit problem (see the following sub-section).

Even if we do not combine different datasets, many housing-related datasets are either spatially incomplete (say, they cover only urban areas in the United States but exclude rural ones), or they have highly varying spatial footprints within one and the same dataset. An example of the latter is HUD Continuum-of-Care program data, see Figure 4.8 (USHUD, n.d.), which sometimes are as big as a whole state and sometimes as small as a mid-sized city (e.g., Fall River, MA).

Other datasets, such as the US EIA's 2022 residential energy consumption survey, initially look impressive as it is based on 18,496 survey respondents – but of its over 100 variables, the finest spatial resolution is that of a state. If we then look at a variable such as the frequency of disconnection notices, we find that this survey is not at all representative, as not a single survey respondent has received such a notice more than once a year – which is inconsistent when compared to the American Housing Survey (2013),6 according to which some 7.8% of all households surveyed received such notices.

### 4.7.1 The Modifiable Area Unit Problem (MAUP)

The modifiable area unit problem (MAUP) is a summary term for two different but related issues when dealing with spatial data. The first is the issue of scale and this is what statisticians call ecological fallacy or the fact that we cannot draw conclusions about specifics from the aggregate. If we know what percentage of the vote a presidential candidate received in a state, then this tells us nothing about how they performed in one of the state's counties. The other aspect of MAUP is unique to spatial data and unfortunately very common. The name "modifiable area" points to the issue of different possible ways to subdivide an area. We will demonstrate this by looking at the various ways different city agencies are carving up New York City.

Figure 4.9 (NYC Open Data, n.d.) illustrates the boundary problem. In a typical housing GIS project, we would compile data from different city agencies. Many of these have their own administrative boundaries; while there are many more, this example shows borough/county boundaries (the city of NYC encompasses five counties), community districts, neighborhood planning areas as defined by the Mayor's office, police precincts, postal ZIP code area boundaries, inclusionary housing areas where zoning has been restricted to support affordable housing, and finally the boundaries around different kinds of zoning (related to but not the same as inclusionary housing). These do not even include school districts or public health planning areas nor many of the special areas such as flood zones, etc.

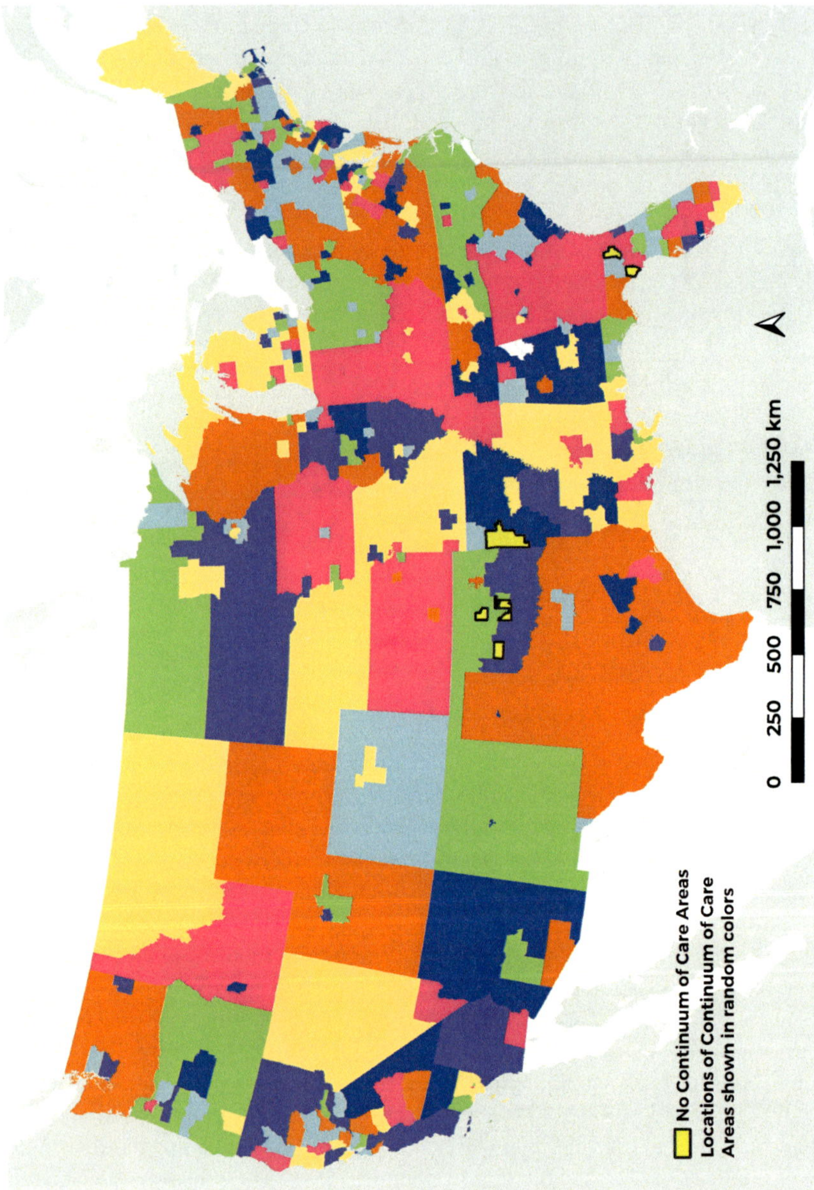


**FIGURE 4.8** HUD continuum-of-care area units

**FIGURE 4.9** Sample snapshot of different administrative boundaries in New York City

Underlying each of these different boundaries are GIS layers that contain a range of descriptors (attributes), which we might want to make use of in our comprehensive analysis. The question then arises, what would happen if any of these city agencies had drawn their boundaries differently? Regardless of what area unit we are looking at, it is the aggregation of individual events (e.g., postal addresses, crime locations, and gerrymandered political boundaries) that then results in aggregate values that *are a function of* how the boundaries are drawn. In other words, if the boundaries have been drawn in a different fashion, the observed aggregate values would be different and would play a different role in our analysis. Research (Openshaw and Taylor, 1979) has shown that, to take an extreme political example, it is possible to redraw electoral boundaries in such a way that in almost every state 100% of all representatives hail from one party only.

This problem would not occur if all our areal boundaries would coincide; if for instance, the postal, electoral, planning, and administrative boundaries (police, school, etc.) would all either coincide or neatly place into each other as many of the US Census boundaries do. The New York City department of City Planning is spending considerable efforts trying to align at least the zoning and planning-related boundaries with those of the Census Bureau in an attempt to minimize the effects of the MAUP. Way ahead of us is France, where most official boundaries align in a neat hierarchical fashion similar to what we discussed about the spine in the US Census hierarchy of area units – except that in France, this applies to postcodes, car license plates, fire and school districts, etc.

## 4.8 NON-CONVENTIONAL DATA SOURCES

In Section 4.6, we described the dearth of data in many aspects of housing research. In this section, we are going to point to a few data sources that are not in the realm of official data but may yet be quite useful.

### 4.8.1 Real Estate Brokerages and Consolidators

The first few are actually quite obvious; as the housing sector (in the United States) is dominated by private businesses, they have an interest in collecting relevant data. Usually, such datasets are proprietary, and many companies seek to maintain their competitive advantage by not disclosing their data. But there are a few exceptions. Redfin is one of the big brokerage companies and was the first one to use a simple online GIS to advertise their properties. They release *weekly, monthly, and quarterly datasets* with several million records each at a spatial resolution of counties and/or metro areas. There are a lot of redundancies in these datasets that require a bit of data wrangling, and the housing researcher would also have to compile their own set of geometries for the counties and metro areas to eventually perform a spatial join (see Chapter 5) to incorporate these datasets into GIS.

Zillow (including its merger with Trulia) and Realtor.com are meta websites that serve real estate agents but also allow individual sellers to list their properties. Their business is to compile the non-standardized records of multiple listing services (MLS) from around the country, resulting in a very comprehensive overview of the residential real estate market. Zillow has both *data download options* as well as an application programmers' interface (API). In addition to owned property data, Zillow also publishes monthly *rental* rates on a per ZIP code areas basis going back to 2014. The Zillow *API allows* developers to query their vast database down to the individual property level, which includes property tax information for almost 150 million properties in the United States. One of their most widely used datasets is a delineation of some *17,000 neighborhood boundaries* in 650 cities which is now hosted by the US EPA.

### 4.8.2 HAZUS Housing Stock Data

A less obvious source of housing data is the Department of Homeland Security's *HAZUS MH program*. An add-on to ArcGIS Desktop, HAZUS is used to model the physical, economic, and social impacts of disasters. The software is of limited use to housing researchers; however, the program comes with extensive datasets needed to estimate potential losses derived from *Homeland Infrastructure Foundation-Level Data* (HIFLD) and the National Structure Inventory (NSI) produced by the US Army Corps of Engineers. The NSI does provide a structure-level representation (as points) of most structures in the U.S., as well as multiple building characteristics including type, occupancy, construction date, building material, utilities connected, etc. However, it is far from perfect and, as with every dataset, should be carefully evaluated to determine if it is suitable for the purpose of any given study.

### 4.8.3 Individual Data Collection; Windshield Survey; Crowdsourced Data

If none of the resources discussed in this chapter fulfill the needs of the housing researcher, then the last resort is to embark on one's own data collection. Sometimes, this is as straightforward as conducting a walking or windshield survey to examine more specific facets of a neighborhood such as

- The age, nature, and condition of the community's available housing
- Infrastructure needs – roads, bridges, streetlights, etc.
- The presence or absence of functioning businesses and industrial facilities
- The location, condition, and use of public spaces
- The amount of activity on the streets at various times of the day, week, or year
- The amount and movement of traffic at various times of day

Windshield surveys require "boots-on-the-ground" but can be a very efficient way of data gathering – especially in a participatory research context. Neighborhood-based researchers can rapidly compile a list of desirable and objectionable characteristics, especially if equipped with mobile phone-based location recording software such as *KoBoToolbox* or *Survey123*. The advantages are the same as for any primary data collection: with complete control over the survey design and collection process, the appropriateness of the data is guaranteed. And if the data collection is performed by locals for locals, then a certain degree of buy-in can be assumed, which helps with respect to quality control. Numerous studies have shown that crowd-sourced geospatial data such as Open Street Map (OSM) is equivalent and sometimes even superior to authoritative data (Zielstra and Zipf, 2010; Zhang and Pfoser, 2019; Jacobs and Mitchell, 2020), which caused, for example, the New York City government to create an agreement with OSM to regularly exchange updates to their respective databases, resulting in one of the best municipal datasets world-wide.

The obvious disadvantage is that the data collection effort will be limited in spatial and temporal scope and cannot be generalized beyond the neighborhood or small city level. Larger surveys become prohibitively expensive, even for experienced and well-funded organizations. The best way to collect housing data on a national scale is to attach the data collection effort to the work of a larger volunteer organization such as the National Low Income Housing Coalition, National Fair Housing Alliance, the National Association of Housing and Redevelopment Officials (NAHRO), the Council for Affordable and Rural Housing, the National Association of Housing Cooperatives, or the National Civic League.

## 4.9 GIS ACTIVITY

In Section 4.3, we presented a figure that illustrates how Housing Affordability has changed over the years and how it also changes geographically. In this section, we are describing the steps it takes to arrive at the maps of Figure 4.4. The topic of this figure is the notion of housing affordability and how it is expressed differently in different parts of the county and developed over time. This starts with a definition of housing affordability. When you google for this term, you will invariably come across the figures of the National Association of Realtors, which by definition covers only (potential) property owners. They release monthly data on a per ZIP code area basis (but have a very restrictive data use policy) that varies mostly because of the month-to-month changes in mortgage interest rates. For the purposes of our example here, we are looking at affordability not just from a homeowner's perspective but every form of tenure.

Our conceptual model takes into consideration rent as a percentage of household income as well as the value of a home in relationship to the owner's income. Neither of these figures are available on a per-household basis. Given that the map in Figure 4.7 covers the whole nation, we decided that county-level data is the appropriate spatial resolution. There are a little over 3,300 counties in the United States, which exhaust the variability that a human observer can handle on a single map. Alternatively, the same data is available at census tract resolution for regional analyses.

Our conceptual model treats renters and homeowners separately. Renters pay their rent on a monthly basis (which is also how it is recorded by the Census Bureau), while homeowners accumulate their assets over the lifetime of their mortgage. Both have additional housing-related expenses such as heating and insurance. But these are complications that do not influence the basic conceptual model. The Census Bureau has been collecting data about mortgage payments but only as of late, making longitudinal analyses impossible. We therefore chose to annualize monthly rents to match annual income values and to spread property values over a 30-year period and then take the annual value as a percentage of the annual income. All the values are using the median values per Census area unit (in our case counties).

For the decadal years 1970–2000, we used data from the NHGIS website, while for 2010 and 2010 we retrieved the raw data from the US Census website. It turns out that for 2010, the US Census Bureau lists ACS 1-year data for only 820 counties, so we had to use the 5-year ACS data for 2010. The universe of counties for the 48 conterminous states varies between 3,008 and 3,011 counties, which has no discernable effect on our maps in Figure 4.4 but constrains a spatio-temporal analysis to only those counties that exist consistently across the five decades.

The ratio of homeowners to renters varies widely across the country. Our calculation of housing affordability therefore weighs the rent burden and homeownership costs according to the percentages of those two categories in each county. After downloading the respective datasets and deriving the base variables for each decade, the calculation of housing affordability is now consistent across the decades. The final step is to calculate the difference in affordability for each county compared to the median value of all counties in the 48 conterminous states. The respective maps depict the difference in shades of red (less affordable than the nationwide median) and blue (more affordable than the nationwide median). The first impression is that housing affordability was much more evenly distributed in the 1970s than in the 2010s. A lot more counties were close to the national median back then than there are now. Affordability was much less an issue in the 1990s than as of late. Particularly striking is the change in the Mountain West where large swaths of the country changed from very affordable to the opposite in only 20 years. The northern Nevada holdout then gave way in 2020 as well.

## NOTES

1. TIGER/Line Shapefiles will be explained in Chapter 5.
2. We will discuss geospatial data formats and storage mechanisms in the following chapter.
3. The actual list is clearly a function of the research question at hand and is likely to differ depending on who is asking it.
4. See, for example, https://files.hudexchange.info/resources/documents/FY-2022-HMIS-Data-Dictionary.pdf.
5. See, for example, https://files.hudexchange.info/resources/documents/FY-2022-HMIS-Data-Dictionary.pdf.
6. Unfortunately, the bi-annual AHS has not asked this question since 2013.

## FURTHER READING

Chu, M, Fenelon, A, Rodriguez, J, et al., 2022. "Development of a Multidimensional Housing and Environmental Quality Index (HEQI): Application to the American Housing Survey". *Environ Health*, 21: 56. doi:10.1186/s12940-022-00866-8.

Devillers, R, and Jeansoulin, R, 2006. *Fundamentals of Spatial Data Quality.* doi:10.1002/9780470612156.

Donnelly, F, 2022. "US Census Data: Concepts and Applications for Supporting Research". *American Library Association Library Technology Reports,* 58: 4.

FEMA, 2022. *Hazus 6.0 Baseline Data Updates. FEMA Factsheet.* https://www.fema.gov/sites/default/files/documents/fema_hazus-6-data-updates-factsheet.pdf, last accessed 4 December 2022.

Goodchild, M, W, Li, and Tong, D, 2022. "Introduction to the Special Issue on Scale and Spatial Analytics". *Journal of Geographical Systems*, 24: 285–289. doi:10.1007/s10109-022-00391-9.

Guptill, S, and Morrison, J, 1995. *Elements of Spatial Data Quality*. Amsterdam: Elsevier.

Hirschman, C, Alba, R, and Farley, R, 2000. The Meaning and Measurement of Race in the U.S. Census: Glimpses into the Future. *Demography*, 37: 381–393.

Missouri Census Data Center, 2022. *Intro to Census Geography, Summary Levels, and GeoIDs.* https://mcdc.missouri.edu/geography/sumlevs/, last accessed April December 2022.

National Institute of Standards and Technology (NIST), 2021. *Compliance FAQs: Federal Information Processing Standards (FIPS).* https://www.nist.gov/standardsgov/compliance-faqs-federal-information-processing-standards-fips, last accessed 4 December 2022.

Sparx Systems, 2022. *Guide to Business Modeling.* https://sparxsystems.com/resources/user-guides/16.0/guidebooks/business-modeling-techniques.pdf, last accessed 4 December 2022.

Uhl, J, Leyk, S, McShane, C, Braswell, A, Connor, D, and Balk, D, 2021. "Fine-Grained, Spatio-Temporal Datasets Measuring 200 Years of Land Development in the United States". *Earth System Science Data*, 13(1):119–153. doi:10.5194/essd-13-119-2021.

USHUD, n.d. *Continuum of Care GIS Tools. HUD Exchange*. Washington, DC: Department of Housing and Urban Development. Online resource available at https://www.hudexchange.info/programs/coc/gis-tools/, last accessed 12 July 23.

Walker, K, 2023. *Analyzing US Census Data*. London: Chapman & Hall.

Wong, D, 2004. The Modifiable Areal Unit Problem (MAUP). In Janelle, D, Warf, B, and Hansen, K (Eds.), *WorldMinds: Geographical Perspectives on 100 Problems*. Dordrecht: Springer. doi:10.1007/978-1-4020-2352-1_93.

## REFERENCES

Albrecht, J, 2007. *Key Concepts and Techniques in GIS*. London: Sage Publications.

Chu, M, Fenelon, A, Rodriguez, J, Zota, and Adamkiewicz, G, 2022. "Development of a Multidimensional Housing and Environmental Quality Index (HEQI): Application to the American Housing Survey". *Environmental Health,* 21: 56. doi:10.1186/s12940-022-00866-8.

Devillers, R, and Jeansoulin, R, 2006. *Fundamentals of Spatial Data Quality*. doi:10.1002/9780470612156.

Donnelly, F, 2022. "US Census Data: Concepts and Applications for Supporting Research". *American Library Association Library Technology Reports,* 58: 4.

FEMA, 2022. "Hazus 6.0 Baseline Data Updates". *FEMA Factsheet*. https://www.fema.gov/sites/default/files/documents/fema_hazus-6-data-updates-factsheet.pdf, last accessed 4 Decmber 2022.

Gold, C, 2016. "Tessellations in GIS: Part I-putting it all together". *Geo-spatial Information Science*, 19(1): 9–25, doi:10.1080/10095020.2016.1146440.

Goodchild, M, Li, W, and Tong, D, 2022. "Introduction to the Special Issue on Scale and Spatial Analytics". *Journal of Geographical Systems*, 24: 285–289. doi:10.1007/s10109-022-00391-9.

Guptill, S, and Morrison, J, 1995. *Elements of Spatial Data Quality*. Amsterdam: Elsevier.

Hirschman, C, Alba, R, and Farley, R, 2000. "The Meaning and Measurement of Race in the U.S. Census: Glimpses into the Future". *Demography*, 37: 381–393.

HUD, and PD&R Edge, 2017. *Defining Housing Affordability*. Washington, DC: U.S. Department of Housing and Urban Development's (HUD's) Office of Policy Development and Research (PD&R). https://www.huduser.gov/portal/pdredge/pdr-edge-featd-article-081417.html, last accessed 4 March 2023.

Jacobs, K, and Mitchell, S, 2020. "OpenStreetMap Quality Assessment Using Unsupervised Machine Learning Methods". *Transactions in GIS*, 24: 1280–1298. doi:10.1111/tgis.12680.

Missouri Census Data Center, 2022. *Intro to Census Geography, Summary Levels, and GeoIDs*. https://mcdc.missouri.edu/geography/sumlevs/, last accessed 12/04/2022.

NYC Open Data, n.d. *NYC Open Datamine*, online resource available at https://opendata.cityofnewyork.us/data/.

Openshaw, S, and Taylor, P, 1979. "A Million or So Correlation Coefficients: Three Experiments on the Modifiable Areal Unit Problem". In Wrigley, N (Ed.), *Statistical Methods in the Spatial Sciences*, pp. 127–144. London: Pion.

Ramasubramanian, L, and Albrecht, J, 2018. *Essential Methods for Planning Practitioners: Skills and Techniques for Data Analysis, Visualization, and Communication*. The Urban Book Series. Cham: Springer. doi:10.1007/978-3-319-68041-5.

Sparx Systems, 2022. *Guide to Business Modeling*. https://sparxsystems.com/resources/user-guides/16.0/guidebooks/business-modeling-techniques.pdf, last accessed 4 December 2022.

Uhl, J, Leyk, S, McShane, C, Braswell, A, Connor, D, and Balk, D, 2021. "Fine-grained, spatio-temporal datasets measuring 200 years of land development in the United States". *Earth System Science Data* 13(1):119–153. doi:10.5194/essd-13-119-2021.

US Census, 2007. *Measuring Overcrowding in Housing*. Report written by the U.S. Department of Housing and Urban Development. Online resource, https://www.census.gov/content/dam/Census/programs-surveys/ahs/publications/Measuring_Overcrowding_in_Hsg.pdf, last accessed 4 March 2023.

US Census, 2020. *Census CQR Frequently Asked Questions (FAQs)*. https://www2.census.gov/programs-surveys/decennial/2020/program-management/cqr/cqr-faqs.pdf, last accessed 16 May 2023.

Walker, K, 2023. *Analyzing US Census Data*. London: Chapman & Hall.

Wong, D, 2004. "The Modifiable Areal Unit Problem (MAUP)". In Janelle, D, Warf, B, and Hansen, K (Eds.), *WorldMinds: Geographical Perspectives on 100 Problems*. Dordrecht, NL: Springer. doi:10.1007/978-1-4020-2352-1_93.

Zhang, L, and Pfoser, D, 2019. "Using OpenStreetMap Point-of-Interest Data to Model Urban Change-A Feasibility Study". *PLoS ONE* 14(2): e0212606. doi:10.1371/journal.pone.0212606.

Zielstra, D, and Zipf, A, 2010. "Quantitative Studies on the Data Quality of OpenStreetMap in Germany". In *Proceedings of the Sixth InternationalConference on Geographic Information Science,* pp. 20–26. Zurich, Switzerland: GIScience, University of Zurich.

# 5 GIS Analysis and Visualization

## 5.1 GIS CORE CONCEPTS

The unique feature that distinguishes GIS from all other software is its ability to help researchers analyze data spatially i.e., to reason about spatial relationships. We have, in the previous chapter, written about the need to organize our data and conveyed our preference for using a database management system (DBMS). Traditional database management systems are not equipped to handle spatial data. End users have to link and integrate spatial information in order to conduct spatial analyses. So, we have to ask, *what is it that makes spatial special*? This chapter cannot replace a formal primer or course about GIS fundamentals. Our goal is instead to provide the thoughtful and serious reader with enough information, so that she can engage in meaningful conversations with GIS specialists. As we discuss the core concepts of GIS, we first have to cover four foundational concepts: (i) coordinate reference systems (CRS), (ii) spatial data types, (iii) spatial operations, and (iv) the geo-relational principle –all of which are unique to GIS, although they can be added to DBMSs. The first three concepts deal with the special spatial nature of GIS-based reasoning and communication; CRS determine where on the surface of this planet our data pertains to, spatial data types are a necessary ingredient to deal with the multi-dimensionality of spatial information, and spatial operations are what allow us to measure distances and directions, and analyze spatial relationships such as adjacency, intersections, or containment. A foundational aspect of GIS is the georelational principle, where every piece of information that we store in a GIS has both a spatial footprint and a set of attributes that describe what we find at the footprint's location.

### 5.1.1 Coordinate Reference Systems

When we work with spatial data, we are usually describing a location on Earth and attempting to describe what we can observe at that precise location (more about this in Section 5.1.2). The location can be a point (e.g., city), or a line (e.g., street), or an area (e.g., county) that has a unique position on the Earth's surface. Coordinate reference systems are used to describe that position (and if the location is larger than a point also the geometric shape of that location). The tricky thing here is the fact that Earth is a spherical object, and that spherical geometry is (i) really hard and (ii) difficult to communicate. Imagine, if you will, attempting to take the entire peel of a juicy orange fruit (a three-dimensional object) and laying it flat on a table and then trying to link a point on the peel to a point on the peeled orange's surface. Whenever we try to transpose a location on the Earth's surface onto a two-dimensional plane, so that we can apply the geometric rules we learned in middle school, we are compromising one geometric characteristic (size, shape, direction, and distance) or the other. Hundreds of different

DOI: 10.1201/9780429243066-5

| Coordinate System | North | West | Zone | Projected |
|---|---|---|---|---|
| Latitude / longitude | 44°58'2.07622" | -103°46'2.07622" | | No |
| Decimal degree | 44.967243 | -103.771556 | | No |
| Universal Transverse Mercator | 4,980,045.51 | 596,875.35 | 13T | Yes |
| SPC 1927 (feet) | 1,024,338.727 | 436,137.9431846 | | Yes |
| SPC 1983 feet | 992,719.62101 | 436,124.6079814 | | Yes |
| SPC 1983 meter | 302,580.76024 | 132,931.8866112 | | Yes |

**FIGURE 5.1** Center of the nation: exploring various coordinate values identifying the geographic center of the United States

coordinate systems have been developed to minimize the distortions and they are all incompatible, i.e., if we combine data that is encoded based on different coordinate systems, we have to translate it from one into the other. A GIS (as well as a spatially enabled DBMS) therefore has to incorporate a library of all the different coordinate systems and has to be able to translate data between the different encodings.

Once we have established what coordinate system to use to describe our positions, we have to decide whether to use two, three, or four values ($x$, $y$, $z$, $t$) to encode a point in two or three spatial dimensions, as well as potentially in time (to capture movements or change). Points are then combined into lines, which are combined into areas, and potentially volumes to describe the spatial phenomenon of interest. Depending on the coordinate system used, we deal either with x and y values or with the latitude and longitude values of spherical geometry. As a rule of thumb, if the coordinate values are small (maximum three digits before the decimal point) then we are dealing with spherical coordinates, whereas if the values are large (in the hundreds of thousands or millions), then we are dealing with coordinates that are projected onto a plane.

Figure 5.1 illustrates the havoc created if the particular coordinate system is not specified. All the locations refer to the same exact position on the Earth's surface. If the coordinate system definition is not provided and instead (wrongly) assumed, the center of the United States may jump around between Wyoming, South Dakota, and Canada, and in the extreme case of assigning unprojected coordinates right on the equator.

### 5.1.2 Spatial Data Types

When we store geographic data, we are either describing features whose locations are given in the form of some geometry or we are describing regular tessellations of space (see Section 5.3.2). In either case, we are dealing with complicated structures that cannot be represented by the data types commonly used in spreadsheet or database programs.

**TABLE 5.1**
**Simple Point Geometries in a CSV File, Stored Together with Attribute Information**

| XY | Lat | Lon | Address | Function | Capacity | Year | Revenue |
|---|---|---|---|---|---|---|---|
| 6274.97, 428422.31 | 38.30742 | −102.8561 | 26 Mall Dr., Town, Zip | Mall | 4,200 | 2006 | 974.2 |

GIS and spatially enabled databases have special data types that allow to store 2- or 3-dimensional coordinates and then combine these into higher-dimensional geometries of variable length. Spatial data types bear some similarity to temporal data types, where we have many different ways to store data and time. Yet, spatial data types are more complicated because of the need to uniquely reference multiple dimensions. To illustrate the point, imagine we are storing city locations as *X*, *Y* values. *X* and *Y* have to be treated not as separate fields but as a tuple (a singular entity consisting of two parts) because if we treat the North and the West values in Table 5.1 independently, then we could sort locations by their "Northness" regardless of their "Westness".

When we store the coordinates that make up the lines of rivers or the boundaries of counties, then there is no way for us to know in advance how many coordinates we need to encode and store a particular county or river. This means that we need data types that allow for variable length of the values stored in them. Alternatively, if we are dealing with spatial phenomena that have no well-defined boundaries, then we can store their spatial footprint in a data structure that is similar to an image – which is yet another spatial data type.

In general, data types designate the amount of memory used to store the data and the internal organization. In addition, data types determine what kind of operation can be performed on the data stored using one type or the other. For instance, when we store spreadsheet data as type character or date, then we cannot perform multiplications on those values. The same is true for spatial data types: once chosen, we are limited to the kind of operations that are applicable for one (spatial) data type of the other. Regardless of what spatial data type we choose the coordinates that we use to store our spatial data are a function of the previously chosen coordinate reference system.

### 5.1.3 Spatial Operations

Spatial operations can be coarsely divided into quantitative and qualitative ones. On the quantitative side, we have measurements of distance and direction, as well as subsequent calculations of areas and volumes. On the qualitative side, we have topological relationships such as inside, outside, touching, and intersecting/crossing. To the uninitiated, these may seem to be fanciful but they are essential for spatial reasoning as well as quality control. If we assume (and this is a rather bold assumption) that all our data is encoded using the same projected coordinate reference system, then we could use drawing programs or CAD systems to calculate distance and direction. But the ability to check whether a particular road crosses a river or a city boundary even if the two do not share a common recorded point is one of the hallmarks of GIS software that no other software is equipped to handle, see Figures 5.2 and 5.3.

FIGURE 5.2 Is the road crossing the river or staying on one side?

### 5.1.4 The Georelational Principle

Throughout this volume, we have talked about geographic entities that are a combination of a spatial footprint and the characteristics that we observe at the location of that footprint. The one-to-one relationship between the two is known as the geo-relational principle. It is mirrored after the basic theorem underlying relational databases, where we link records in one table with records in another table. Applied to geographic data, this link is now between a record describing a specific geometry at a unique location and another record in a table of non-spatial attributes. This requires each table to have a field containing a unique value (primary key) and there to be one and only one corresponding record containing the same key values in the two tables that form the relationship.

**FIGURE 5.3** Encoding a lake on an island that is inside a lake that is on an island that is inside a lake. A highly nested topological relationship found in Yathkyed Lake in Nunavut, Canada

Once established, the georelational principle allows for querying the database either by location (usually in the form of an interactive map user interface) or by field values as we would commonly do in a database or a spreadsheet. The simultaneous exploration of geospatial data using either the map or the table interface is extremely powerful. But before we get into the (exploratory) data analysis possibilities afforded by GIS, we will have a look at what contributes to the popularity of GIS. In the case depicted in Figure 5.4, a table query resulted in the city of Thiruvananthapuram, India, being selected and then being marked in yellow on the map. Alternatively, we could select any of the feature geometries on the map to then display its geographic attributes in the table that is linked to the geometries by the georelational principle.

## 5.2 GIS MODELS

Until now, we have been very vague with respect to the geometries used to position the objects of our inquiry on the Earth's surface. We gave examples of zero-, one-, and two-dimensional features and mentioned that traditionally, the geometries are stored separately from the attributes, where we characterize the nature of the things we want to reason about. Historically, this separation made a lot of sense because we could continue to work with spreadsheets and database tables for the non-geometric components and kept the specialized geometry descriptions (as well as the coordinate system information) separate. Another advantage of this separation is that we don't have to accommodate for the many different data types as part of our table definitions.

### 5.2.1 Spatial Data Formats

The easiest way to transition from a simple table is if the spatial reference is just a point. In that case, we may remain with a comma separated value file, where we put the point information in quotes, which allows us to store *x*, *y*, *lat*/*lon*, or even address information.

The spatial information depicted in Figure 5.4 is overly redundant, although it is common in municipal data to store the same information in multiple ways to accommodate the needs of different audiences. Things get a little more complicated, the moment our spatial reference is a linear or areal object, not to mention non-simple geometries. This is where we encounter the historic split into data formats that separate out the geometries and more modern representations that accommodate variable length fields.

The most widely used format that follows the logic of the geo-relational principle is what is misleading called the shapefile. It is misleading because a shapefile is actually a combination of at a minimum three and possibly as many as seven different files that have to be co-located in the same directory or folder and all have the same name but different file extensions. Because the shapefile is a combination of files, they are usually exchanged in the form of a .zip archive. What is confusing about the name is that one of these required components of a shapefile is a file with the extension .shp, which contains the geometry information. The other two required files are .dbf, where the attribute information is stored, and .shx, which implements the georelational principle by linking each record in the .shp file with its counterpart in the .dbf file. There are other geospatial file formats that implement the georelational principle by the same vendor (Esri) and by others but the shapefile is by far the most common one. It has been around for over 30 years and has a number of disadvantages, including but not limited to:

- Attributes are stored in a dBase file, which hails from the early 1980s and carries the limitation of the early MS-DOS operating system, namely very few data types and severely limited variable name conventions
- Geometry types are separated, i.e., points, and lines, and areas have to be stored in different shapefiles
- There is no way to store topological relationships

The shapefile format *used to be* the default in many different GIS but the above-mentioned disadvantages led to the development of a multitude of more flexible GIS data formats. Beyond the realm of GIS, markup languages are providing the basis for a number of geospatial formats that can be encoded as ASCII files (similar to the original .csv format) but now allowing to encode geometry information in the form of long strings. Both the original keyhole (.kml) as well as the geography markup language (.gml) fall into this category. Among web developers, the Javascript Object Notation (JSON) is widely popular and has spawned geospatial variants in the form of geoJSON and topoJSON. A decade ago, these would have been considered unwieldy because their plain ASCII storage causes these files to be rather

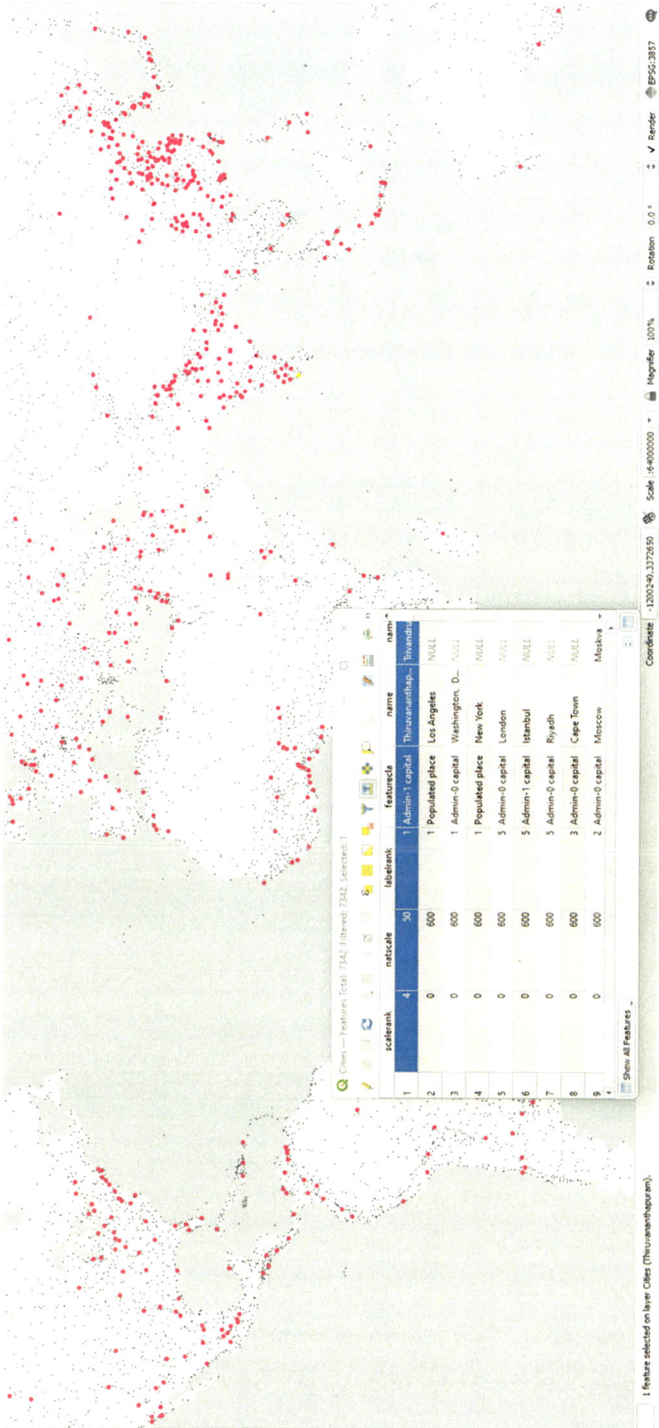


**FIGURE 5.4** Showing a query by location vs. a query by attribute

voluminous. However, in the age of "Big Data", this does not seem to be an issue anymore, and the easy readability and their similarity to data formats beyond GIS makes them now very popular data exchange formats for geospatial data. Most open data repositories now offer geoJSON and GML as a download option.

An additional advantage of these formats is that they can be loaded into a simple text editor and parsed by non-expert GIS users. For local storage and efficient analysis, however, housing researchers should adopt a spatially enabled database. Larger organizations will probably already have their in-house DBMS, which can be spatially enabled (for free, if the DB is open-source). Smaller organizations or individual researchers are better served with personal databases that implement a DBMS in a single file. In the 1990s, this was exemplified by MS-Access but now we have specialized (and standardized) geospatial databases like SpatiaLite and building on that the GeoPackage format, see Figure 5.5. SpatiaLite is an extension of SQLite, an open source database that is built into every mobile phone, many operating systems, and appliances. The GeoPackage "is an open, standards-based, platform-independent, portable, self-describing, compact format for transferring geospatial information" (OGC, 2022).

It is now the default storage format for the widely used free and open-source software Quantum GIS (*aka* QGIS) and is suitable for all but the largest GIS implementations (for which a full-fledged DBMS is a must). Because .gpkg files implement a whole database in a single file, they are easy to share or archive. GeoPackages implement a multitude of common geometries including

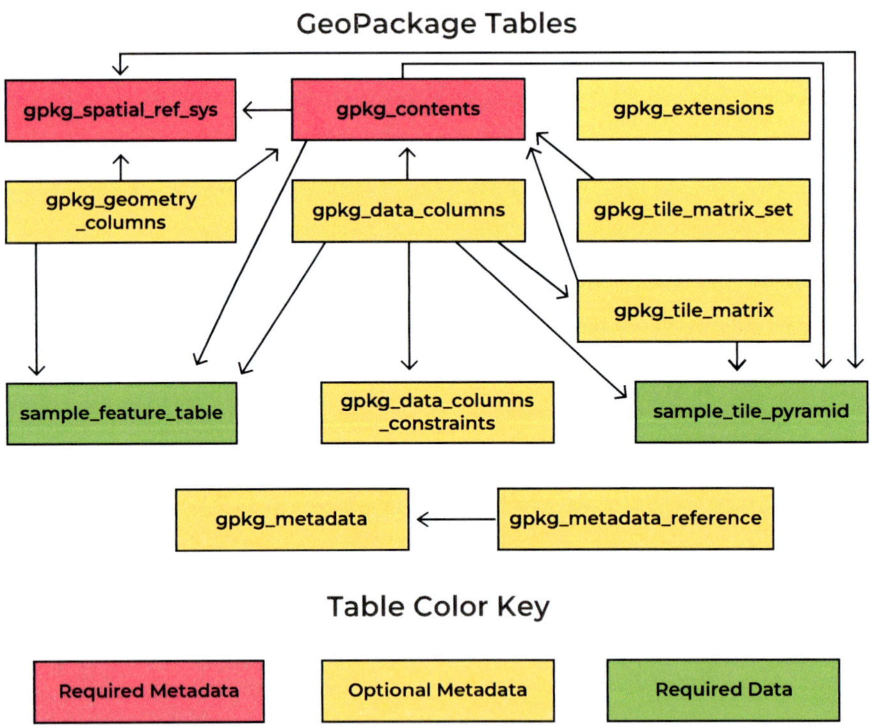


**FIGURE 5.5** Required and optional components (tables) of a GeoPackage

a. Vector feature data
b. Imagery tile matrix data
c. Raster map tile matric sets
d. non-spatial tabular data, and
e. metadata that describes other stored data

Items (a) through (c) will be discussed in the following sub-sections.

### 5.2.2 Spatial Data Models

The first three bullet points in the content list of a GeoPackage required some elucidation, as they describe formalizations of descriptions of space that are common to *all* GIS and as such reach beyond the scope of GeoPackages alone. So far, we have always referred to points, lines, or areas as the spatial footprint of our objects of interest. In the world of Housing GIS, these types of geometries are by far the most common; in the world of GIS more generally, they are referred to as vector features. The term "feature" is used whenever we are dealing with something that has a well-defined boundary and in addition to the geometric description of those boundaries implements the geo-relational principles by adding non-spatial attributes. The term "vector" derives from the mathematical origins of encoding the *boundaries* of features. Contrary to popular conceptualizations, an area (or in GIS terms, a polygon) is not described by what is inside the area but by its boundary. The boundary of an area is made up of a minimum of three (but potentially thousands of) lines. Everything inside the boundary is taken to be uniform; there is no further differentiation of such an area, as this would require another boundary – as in the island-in-a-lake example depicted in Figure 5.3. The lines that make up the area boundary are again defined by their respective boundaries: the start and end points of each straight line. Zero-dimensional points also have a boundary themselves. So, everything in the world of vector features boils down to a collection of points, which are defined by their position relative to the origin of the coordinate system. The imaginary line from that origin to the position of a point is called a vector – hence the name vector feature and by extension Vector GIS.

Complementary to the way of conceptualizing entities in space by their respective boundaries is the notion of a field. Fields are well-known in the physical sciences: electric, magnetic, gravitational, etc. fields. What characterizes fields is their lack of boundaries. They represent phenomena that are clearly discernible but hard to fix in space. Most aspects of nature fall into this category: where is the beginning or end of a mountain, a (natural) forest, a coastline (don't forget the tides)?[1] As there are no boundaries, traditional vector geometries would be useless for describing such phenomena. GIScientists solved this conundrum by describing space, known as a raster, rather than objects in space. The term is of German origin and would typically be translated as grid. A raster/grid divides a study area into uniformly shaped and sized areas: triangles, squares, or hexagons, with square being by far the most common tessellation – although hexagons are becoming more popular as of late.

Both the remote sensing community and the GIS community have been inventing this data model in parallel. There are lots of similarities between the images in remote sensing (which, contrary to images taken with a camera or by a desktop scanner, are also georeferenced) and the grids used in raster GIS. The rationale to use rasters could either be the application (where there are few or no discernable boundaries) or the data capture instrument (a camera or similarly working instrument). Where vector features have a scale (the smallest element represented), raster datasets have a resolution (the size of each grid cell in units on the ellipsoidal surface representing Earth). For a given extent, the finer the resolution, the more cell values have to be recorded; this causes raster datasets to be significantly larger than vector files (recall that in the vector world, we record nothing about the inside of areas, whereas in the raster world, each cell has to be recorded/stored). To avoid having to work with very large files, raster, and image files (also referred to by their mathematical name "matrices") are indexed by tile pyramids (see Figure 5.6). A tile pyramid is a table that links to multiple resolutions of the same original raster layer. This is another reason to use databases because in addition to multiple vector files, a database can also store each raster and each of the multiple resolutions of a raster as separate tables that can be linked by yet another table.

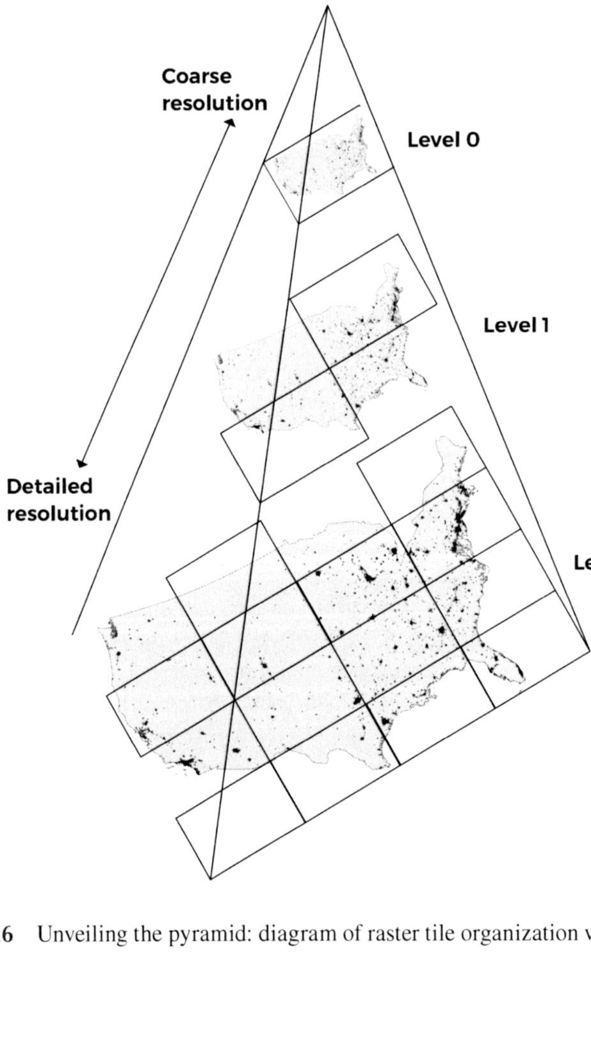


**FIGURE 5.6** Unveiling the pyramid: diagram of raster tile organization with a hierarchical structure

## 5.3 BASIC GIS ANALYSIS OPERATIONS

Until now, we have seen two main reasons to use GIS: (i) to compile a range of datasets in a database, where we use the spatial component to index and link them by location, and (ii) to use the map interface to explore spatial relationships visually, that is to use visual cues as prompts for subsequent analysis. Many novice GIS users combine datasets in GIS to overlay them visually and then use their perceptual prowess to determine relationships between features in different layers. Combining different pieces of information on a map is a good first step – but it needs to be followed by a second, where we use the power of GIS to analytically support (or reject) our observations.

What qualifies an operation to be analytical? GIScientists have an interesting perspective on this. They distinguish between (simple) queries that retrieve an existing item from a database and analytical operations, which create something new (that did not exist in the database). The boundary between the two is fuzzy but we are on the safe side if we just check whether the result of an operation is a new geospatial dataset. If yes, then this operation falls into the analysis category. As this section is about basic GIS analysis operations only, we can now separate them into two sets of operations: horizontal and vertical. Horizontal operations usually involve only one layer and our perspective is outward bound from our object(s) of interest; they are also referred to as neighborhood operations. Vertical operations look across multiple layers and seek to determine which objects or characteristics spatially coincide; they are also referred to as overlay operations. See Figure 5.7 as a visual representation of the difference between neighborhood and overlay operations.

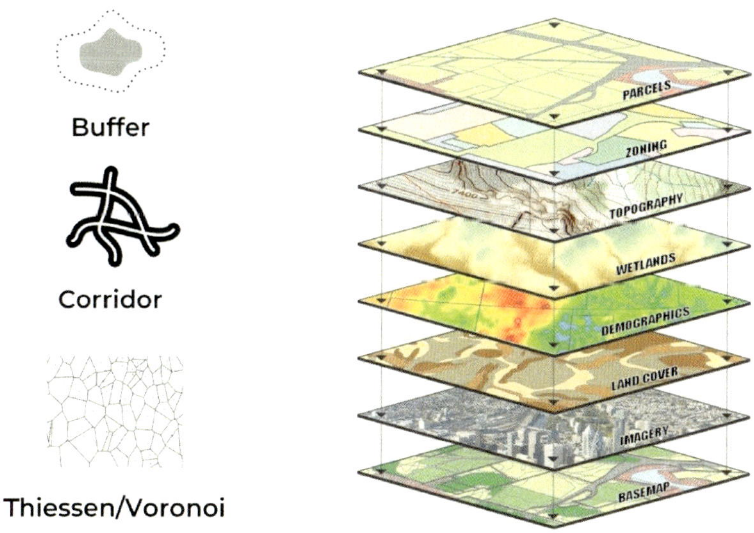


**FIGURE 5.7** Neighborhood vs. overlay: diagram comparing buffer, corridor, and Thiessen/Voronoi operations illustrating spatial relationships and analysis techniques

### 5.3.1 Neighborhood Operations

Neighborhood operations take a location of interest (in the case of a raster representation) or the spatial footprint of a feature of interest and then have the user define an extent (the neighborhood) around it. That definition may be a simple number (units distance from the location of interest) or a heuristic where the definition of a neighborhood is a function of some attribute. Neighborhood operations then either just define a new set of features that delineate the boundaries of the neighborhood (or raster cells contained in each neighborhood) or they perform calculations on the features/raster cells within a neighborhood. This is then repeated for all features in a layer or all cells in a raster dataset respectively.

By far the most common neighborhood operation is the buffer operation, which results in a new dataset that contains all the buffers around the input features or cells. In its simplest form, the user specifies a distance, say 1,000 feet, and the GIS will create a new layer with areas of a 1,000 feet radius around the input features (e.g., bus stops). A prominent use of buffer zones in many cities is the legal requirement to identify zones around schools, where liquor (or cigarettes) may not be sold.

Another common neighborhood operation is the generation of so-called Thiessen polygons or Voronoi diagrams.[2] The input to this operation always consists of points (schools, fire stations, hospitals, etc.). Now imagine, we are simultaneously buffering all the point features (raster cells) by ever increasing distances until the boundaries of our buffer regions meet. Where they meet, we stop, but where there is still a gap, we continue our ever increasing buffer distances, see Figure 5.8 (NYC Open Data, n.d.). The process stops when there is no space left and the study area has been completely tessellated. Each input point is now surrounded by polygons that define the point's catchment area, where every location inside the catchment area is closer to the original point than to any other point. Such delineation of catchment areas is of obvious interest to every spatially aware social scientist.

So far, we assumed that as we define the distance to the input location, the new boundary is measured "as-the-crow-flies", i.e., without incorporating any obstacles. This is acceptable for phenomena that spread continuously such as noise or an air pollutant, but it is unsatisfactory for measures of accessibility. Social scientists may be more interested in taking a particular distance measure (which could also be scaled by time or safety) and then applying it along a network representing streets, or sidewalks, or transit lines.

### 5.3.2 Overlay Operations

Useful as they are, neighborhood operations are by far outweighed by overlay operations. As a matter of fact, for many, the whole purpose of GIS is to perform spatial overlays. This is problematic because, although the set of all different overlay operations is definitely very important and arguably makes up over 50% of all analytical GIS operations in practice, there is a world of difference between the visual overlay we discussed at the beginning of this Section 5.3 and the analytical overlay here. Let's keep in mind that analytical GIS operations *always result in new data*, not just a new map but also a new dataset that can be queried and quantitatively analyzed. Visual overlays, i.e., just displaying multiple layers in a map, is a good way to trigger research questions – but not to answer them. Rigorous housing GIS research requires

**FIGURE 5.8** Fire station influence: map of Manhattan with Thiessen polygons suggesting coverage areas of each fire station

us to perform analytical overlays and understand what happens under the hood when we instruct the GIS to run one form of overlay analysis or the other.

This is perhaps best illustrated by thinking back to the MAUP (Section 4.7.1 in Chapter 4). If we want to learn about the median age of housing stock in a ZIP code area, then we have to reason about datasets that have different spatial footprints. One may be a high-resolution property database, while the other is a definition of neighborhoods (ZIP code areas may be replaced with Zillow real estate neighborhoods, or Midwestern aldermanic districts). In any case, we may now use the larger area units as a cookie cutter to aggregate the building age values. Next, we want to combine this with the fire department's incident or inspection data. The MAUP occurs whenever boundaries of one analysis unit do not coincide with the boundaries of another analysis unit. We are now overlaying the ZIP codes area data with the fire district data. In an ideal case, we can always go back to the fine-grained property data and link the building age to the inspection or incidence records at the property address level. In that case, we do not have a MAUP issue. The classic GIS overlay situation, however, is when we are trying to combine the ZIP code level data with the fire district data. In this case, we are looking at combining the two different geometries with the two different attribute datasets. Analytical overlay operations may involve point, line, area, as well as raster datasets. And as we combine different geometries, we have to look at their topological relationships to perform the analyses. What is happening under the hood is a sequence of steps that create new geometries and then subsequently new attribute records to match those new geometries. So, let's go through these step-by-step.

Overlay operations were described as working vertically, i.e., for each location, we ask what is happening here (in this layer) and what is happening at the same location in other layers. We are comparing spatially coinciding values with each other. This in turn means that for each location, we need to look up whether we are in one feature or the other (this is a lot easier in the raster world, where we do not have feature boundaries and hence can easily compare coinciding raster cells/locations). We mentioned earlier that in vector GIS, we don't say anything about the interior of polygons – they are defined by their boundaries. This in turn means that in an overlay operation, we need to determine whether we are outside, on the boundary or inside a particular feature. We then determine the same for the features in the other layer(s) and then create new features that inherit the characteristics from their respective parent features in the input datasets. The first step is to see where we are with respect to each and every feature in our input datasets.

We then determine the same for the features in the other layer(s) and then create new features that inherit the characteristics from their respective parent features in the input datasets. The first step is to see where we are with respect to each and every feature in our input datasets. We compare outsides, on-the-boundaries, and insides of all elements and thereby determine which ones are coincident at what location. The three qualitative options are defined by the topological relationships of the participating features (it does not matter how far inside or outside we are). Several researchers working with Max Egenhofer at the University of Maine and Eliseo Clementini at the University of Aquila, Italy, developed the mathematical proofs to exhaustively formalize all possible topological relationships between the boundaries of point, line, and area objects in the early 1990s.

For each of the seven groups depicted in Figure 5.9, there is a different GIS overlay operation. Each of these operations is complex; the software needs to determine what type of geometries are involved and then perform complicated geometry calculations for each and every feature of the respective layers. This is computing-intensive and can still take hours on large datasets. The results are new, and in most cases smaller geometries than in the input layers. Multiple consecutive overlay operations result in so many small geometries that they have to be followed up by some form of reclassification that is based on the most useful combination of attributes for the research question at hand.

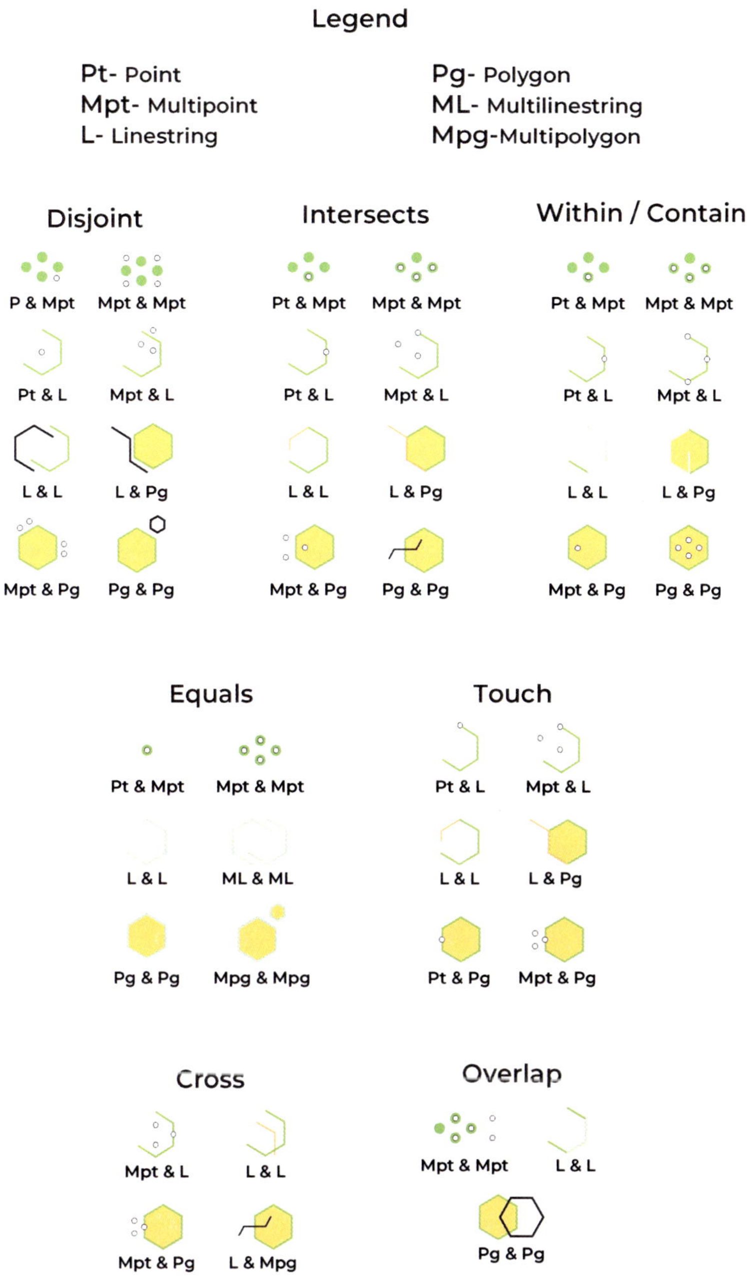


**FIGURE 5.9** Exhaustive enumeration of topological relationships between 0-, 1-, and 2-dimensional geometries

Which brings us to the other side of the georelational principle? Each overlay operation involves not only the geometries but combines attributes as well. The effort we put into data cleaning and conceptual model development in Chapter 4 now really pays off because the more succinct the inputs to the overlay operations are, the easier it is to now instruct the GIS how to combine attribute values: should they be added, averaged, or reapportioned as a function of size of the areas? Overlay operations are indeed very powerful and may even be the essence of GIS.[3] With this power comes the responsibility of the housing researcher to understand the difference between the seven types of overlay operations and the need to develop a conceptual model that guides us in the choice of which operation to apply.

Between the neighborhood and overlay operations, we covered around 70% of analytical GIS operations, housing researchers are going to apply on a regular basis. Before we deal with the remaining 30%, let us have a look at how the basic analytical operations are used by housing researchers in a set of typical examples.

### 5.3.3 From Simple GIS Operations to Workflows

In Section 4.5 we discussed conceptual models as the foundation for a database schema. This is a good and necessary step, for if we don't have our data in place and properly organized, then there is nothing that we can apply our GIS operations to. But housing research is typically more complicated than just applying one GIS operation or the other.

At a high level, a typical GIS workflow would consist of these nine steps:

1. Define research objectives: Clearly outline the goals of the housing policy research, such as identifying areas with a high concentration of affordable housing or analyzing the impact of zoning regulations on housing development.
2. Collect data: Gather relevant data from various sources, such as census data, housing market data, zoning regulations, and land use data. This data will be used to create GIS layers and perform spatial analysis.
3. Data preparation: Clean and preprocess the collected data to ensure its accuracy and consistency. This may involve geocoding addresses, converting data formats, and standardizing attribute information.
4. Create GIS layers: Import the cleaned data into a GIS software and create layers representing different aspects of the housing policy research, such as housing prices, zoning regulations, and population density.
5. Perform spatial analysis: Use GIS tools and techniques to analyze the relationships between different layers and identify patterns or trends. For example, you might use spatial overlay analysis to determine the areas with the highest concentration of affordable housing or buffer analysis to identify the impact of zoning regulations on housing development.
6. Visualize results: Create maps and other visualizations to effectively communicate the results of the spatial analysis. This may include thematic maps, heat maps, or 3D visualizations.
7. Interpret findings: Analyze the results of the spatial analysis and draw conclusions about the housing policy research objectives. This may involve identifying areas in need of affordable housing development or recommending changes to zoning regulations to promote housing diversity.

8. Communicate results: Share the findings of the housing policy research with stakeholders, such as policymakers, housing developers, and community members. This may involve creating reports, presentations, or interactive web maps to effectively communicate the results and support data-driven decision-making.
9. Monitor and evaluate: Continuously monitor the housing market and policy changes to evaluate the effectiveness of the research and make adjustments as needed. This may involve updating the GIS layers, conducting additional spatial analysis, or refining the research objectives.

If our research question is to analyze the effect of changing zoning rules to allow for accessory dwelling units (ADUs), then step 5 above can be further broken into this sequence of GIS operations:

1. Identify zoning layers: Start by identifying the zoning layers in your housing database that are relevant to the research question. This may include layers representing current zoning regulations, land use, and existing housing stock.
2. Create a new zoning scenario layer: Make a copy of the current zoning layer and modify it to reflect the proposed changes, such as allowing ADUs in specific zones or relaxing density restrictions.
3. Overlay analysis: Perform an overlay analysis to identify parcels that would be affected by the zoning changes. This involves overlaying the new zoning scenario layer on top of the existing land use and housing stock layers to identify parcels where ADUs would now be allowed.
4. Calculate potential ADU capacity: For each affected parcel, calculate the potential number of ADUs that could be added based on the new zoning rules. This may involve considering factors such as lot size, setbacks, and maximum building height.
5. Summarize potential ADU capacity by zone: Aggregate the potential ADU capacity calculated in the previous step by zoning category or neighborhood to get a better understanding of the overall impact of the zoning changes on ADU development.
6. Analyze the impact on housing affordability: Assess the potential impact of the increased ADU capacity on housing affordability in the affected areas. This may involve comparing the potential ADU capacity to current housing demand, analyzing the potential impact on housing prices, or estimating the number of affordable units that could be created through ADU development.
7. Assess the impact on infrastructure and services: Analyze the potential impact of the increased ADU capacity on local infrastructure and services, such as transportation, schools, and utilities. This may involve using GIS tools to estimate the additional demand for these services and identifying areas where upgrades or expansions may be needed.
8. Visualize the results: Create maps and other visualizations to effectively communicate the results of the analysis. This may include thematic maps showing the potential ADU capacity by zone or neighborhood, with heat maps illustrating the impact on housing affordability, or 3D visualizations depicting the potential changes to the built environment.

Both of the above lists are fairly generic. The first one applies to virtually all GIS projects, regardless of whether they are in ecology, crime analysis, or housing research. The second list is more specific to our application area but still generic enough to be replicated, say for each neighborhood in a city – with slightly varying parameters as our requirements change from one location to another. It is worthwhile mentioning that the operations themselves are very basic; their impact derives from the repeated application of the same small set of basic operations to intermediate outcomes. If we can save the sequence of processing steps as a model that can be executed with a single click, then we (i) avoid the tedium of repeated the same steps again and again, (ii) ascertain that when we run the model again it can be compared with previous model runs because the steps are guaranteed to be the same, and (iii) we can share this model with a colleague. In information programming terms, this would be called creating a function. In the world of GIS, this model creation is referred to as geoprocessing (a term coined by the company Esri) or just plain processing (in the world of free and open source GIS).

The reason we began this section with a nod toward our discussion of conceptual models in Section 4.5 is that we should treat the development of such processing workflows as the other side of the same conceptual modeling coin. One of the authors of this volume has built his career on the development of tools for such workflow modeling. Simple models can be built with GIS-internal tools but complex models that link to larger institutional (and non-spatial) workflows would benefit from using either the Unified Modeling Language (UML) or the software that implements the standards of the Business Process Modeling Notation (BPMN). However, regardless of whether we sketch out our workflow on the back of a napkin (not a bad idea!) or using a formalized language, the development of a workflow sequence (i) helps to clarify in one's mind what exactly it is we are trying to accomplish with our GIS work, (ii) helps us to document our workflows both for the sake of communicating it in a final report as well as to build institutional knowledge, and (iii) develop a small library of standardized workflow models that are unique to the enterprise we are working for and can be deployed with the push of a button to anyone with a barebone knowledge of GIS.

### 5.3.4 Basic GIS Functionality in Housing Policy Research

Section 5.5 will provide some in-depth examples for GIS use in housing policy research. This subsection is a prelude to provide the reader with a few practical examples of the otherwise rather abstract and technical discussion of basic GIS analysis operations. We will illustrate the application of neighborhood and overlay operations with two commonly asked questions: (i) is there a relationship between building permits and gentrification, and (ii) are rents higher or lower near transit stops? The first question can be answered with overlay operations only, while the second question requires a combination of overlay and neighborhood operations.

The first question is also a fine example for the importance of conceptual models because depending on how we conceptualize the term gentrification, we would try to capture this phenomenon with a range of different variables. Even something as plain as building permits deserves a little further scrutiny because permits for new construction typically do not cause displacement (unless it is preceded by demolition), while building alterations often require tenants to vacate at least temporarily. In an aspatial world, we would just look for the respective values of whatever variables we found to be representative of our research question and then look at trends on a city-wide or state-wide scale to determine whether there is a correlation between the number of permits and the gentrification indicators. With GIS, however, we are aiming to capture the variations in space. Where do we have how many buildings permits and where do we observe what gentrification indicator values, be they rental price increases above the regional average or the percentage of people who did not live in a neighborhood 5 years ago? Each variable becomes a GIS layer that allows us to depict the local or regional differences. Assuming that the gentrification variables are combined into a summary indicator, we can then perform an overlay between the building permits data (typically point data that we could summarize to the level of area units that we measure gentrification in) and the gentrification layer. This spatial perspective will then provide us with evidence for where there is the presumed relationship and where there is not.

The second question takes a horizontal perspective, where again, we have to consult our conceptual model to determine what "near transit stops" means. What mode of transit should be included and how far do we anticipate the influence to reach? In theory, we might even do without any preconceived notion of horizontal reach because in a perfect world, we would need to only map the spread of rental rates and if there is a relationship, then we should observe hot or cold spots (troughs and peaks in a 3-dimensional representation) wherever there is a transit stop. But chances are that the spatial relationship between our two observables varies across the study area and so we typically define catchment areas around each transit stop (either simple buffers as-the-crow-flies or along a road network depicting temporal isolines) and then compare the average rents inside the catchment areas with those outside.

In Section 5.5, we will delve a little deeper into the range of GIS analyses that are available to housing policy researchers. But before we go there, we need to discuss the role of visual communication that accompanies any GIS analysis.

## 5.4 GIS FOR MAPPING AND VISUALIZATION

One of the main attractions of GIS is its ability to engage the housing policy researcher through its interactive map-based user interface. It is this visual representation and the opportunity to interactively explore spatial relationships on a map that sells GIS to larger audiences. Visualization occurs at all stages of the GIS process. Whenever we receive a new dataset, we should look at it both from a descriptive statistics perspective as well as display the data on a map. In both instances, a cursory (but purposeful) look at the data will give us clues about their usefulness; but what is unique about the map is that it is prone to draw us into exploring spatial relationships. The map will provide us with situational context and prompt us to look for patterns. This is built into us humans; we may actually detect patterns that turn out not to be statistically relevant – but this is what the analytical part of GIS is for. Most people looking at a map will try to reconcile what is displayed with what they know about the place. The mere display on a map will either confirm what we know or will invite questions about whatever surprises us, see Figure 5.10 (US Census, n.d.; NYC Open Data, n.d.).

This process of visually making sense of the data should be done for each dataset individually, and then by looking at the relationships between the different datasets. Part of the mythos of GIS is that each dataset becomes its own map layer and that we can stack map layers on top of each other to then visually explore the relationships

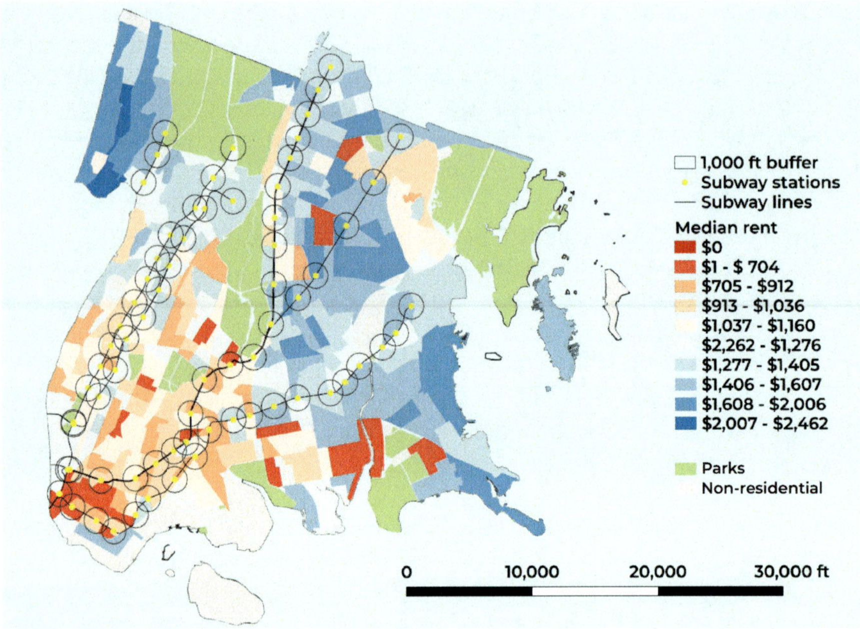


**FIGURE 5.10** Rent and transit nexus: GIS map showcasing the relationship between median rent and proximity to subway stations using multiple layers

among the map elements *across* the layers. This takes us back to the notion of conceptual models discussed in Section 4.5 as well as the basic GIS examples at the end of the previous section. Which data points coincide spatially or are in close vicinity to each other? Is there a relationship between building permits and gentrification? Are rents higher or lower near transit stops? The visual exploration will again help us to generate research questions and to check our initial assumptions (which will have to be confirmed using the analytical methods of Sections 5.3 and 5.6). The ability to jump back and forth between the table and the map interface and to have these linked through the georelational principle is one of the big selling points for GIS in housing research.

### 5.4.1 Tapestry Data

One of the best examples for putting our own data into context and then applying spatial reasoning is Esri's tapestry segmentation data, a well-developed example of geodemographics that identifies 67 different spatialized market segments. Using data clustering and data mining techniques (partially discussed in Section 5.6), Esri delineated contiguous areas (which they call neighborhoods) throughout the United States, where the resident population falls into one of the 67 euphemized demographic categories listed in Table A.7, located in the appendix. Now, a serious housing researcher will compile contextual data herself rather than relying on the marketing-oriented tapestry segmentation data. However, for GIS students, this represents an excellent example of how to make sense of the housing geography of a place – especially if one is not a local. See Figure 5.11 for a map made with Tapestry Data.

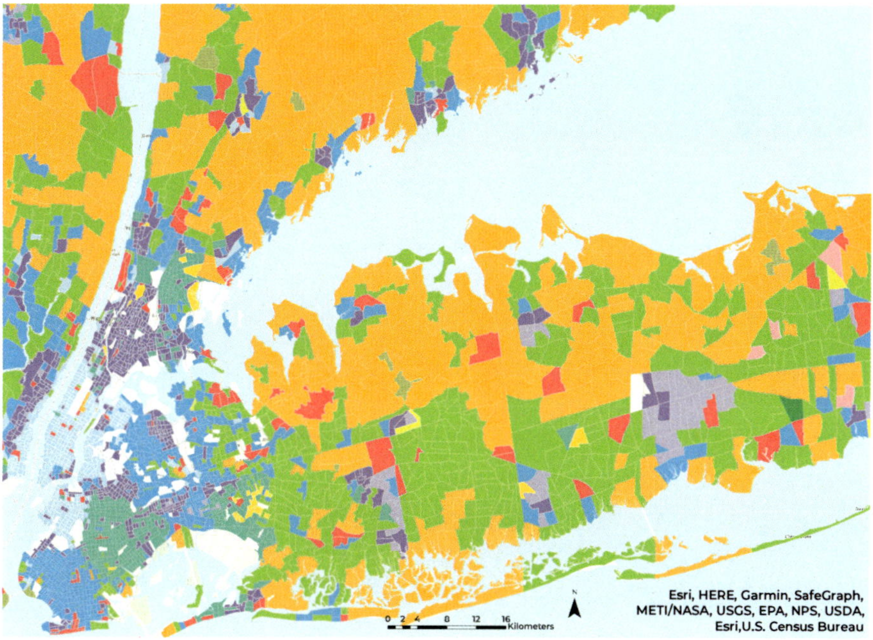


**FIGURE 5.11** Market segment clusters in the Metro NY tapestry data

### 5.4.2 Data and Information Visualization

This volume is based on the premise that spatial differentiation matters and all the maps in this chapter so far are an illustration of the advantages of GIS when it comes to analysis. But as discussed in Section 5.2 policies need to be communicated and maps are a natural ally of the housing researcher – if deployed conscientiously.

Take Baltimore's online *Community Development Map* (CoDeMap), for instance. It visualizes housing needs in the city, neighborhood by neighborhood. CoDeMap is a central point of access for the housing department's numerous databases with everything from citation data to a property's permit history. It has evolved from a housing code enforcement tool to a platform that provides insights into housing, community development, and property datasets at the citywide, neighborhood, block, and parcel levels. It is this double function of serving both inward-facing city employees to link data across different repositories to answer specific questions, and serving the public that displays the power of GIS.

On the inward-facing side, CoDeMap can display a census block or parcel level to reveal foreclosures, open work orders, outstanding violations, property types, vacancies, ownership types, and more. Having all key data in one place also allows staff from other city departments to *see* and understand housing policies. Much of this is now shared with nonprofit organizations, neighborhood associations, and developers, who have received free training sessions that allow them to explore the riches and help to create an equal playing field when it comes to discussing new development plans. GIS visualizations (maps as well as the storyboards of the following subsection) are an immensely effective communication, discussion, and public engagement tool, Figure 5.12 (Baltimore DHCD, 2023), which highlights Baltimore's Community Development Map.

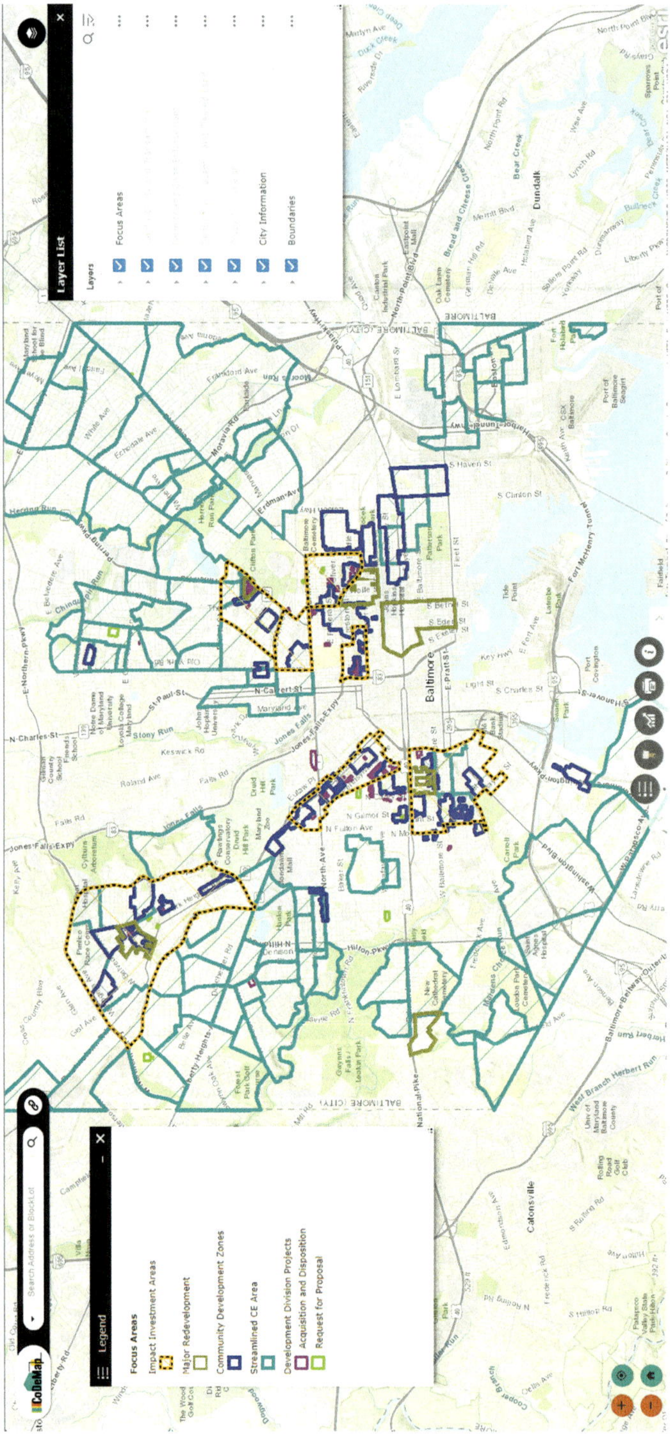


**FIGURE 5.12** Community development map highlighting areas of redevelopment, development zones, impact areas, streamlined regions, and ongoing projects

### 5.4.3 Communication to Different Audiences

Where the previous subsection concentrated on visualizing the co-location of different aspects of a planning decision in a desktop environment, we are now discussing examples of taking interactive GIS displays to the Web. Public outreach is a legal requirement for virtually all housing policy decisions. Figure 5.13 (Chester, 2023) could hail from a traditional static local planning department webpage. But this is just the luring entry point to a website that then engages the visitor with its ability to query the system based on their own home address (Figure 5.14, (Bucks, 2023)). It is easy to engage citizens if they are given the means to find out what is happening in their vicinity. Northern Kentucky's *Link GIS website* rivals any popular social media site with its storymaps, a mashup of text, background photos, videos, and interactive maps that we introduced in Section 3.6. By translating each (GIS) project into an *engaging story*, Link-GIS keeps justifying its existence to

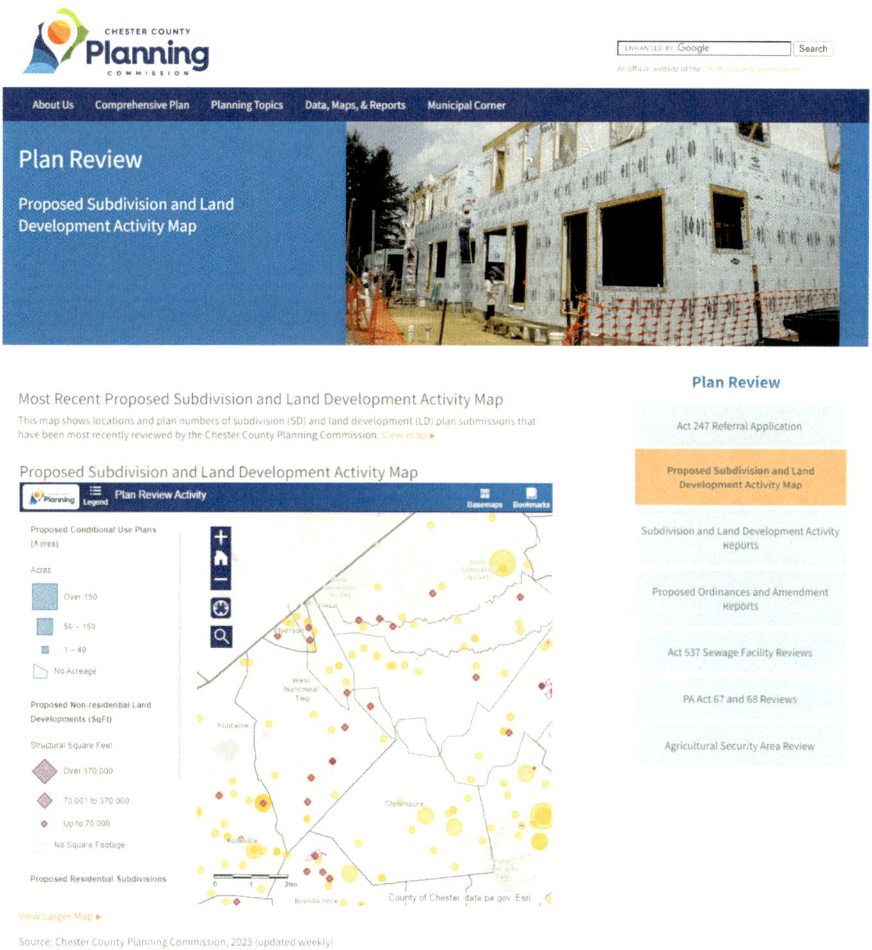


**FIGURE 5.13** An attractive entry point to an online GIS

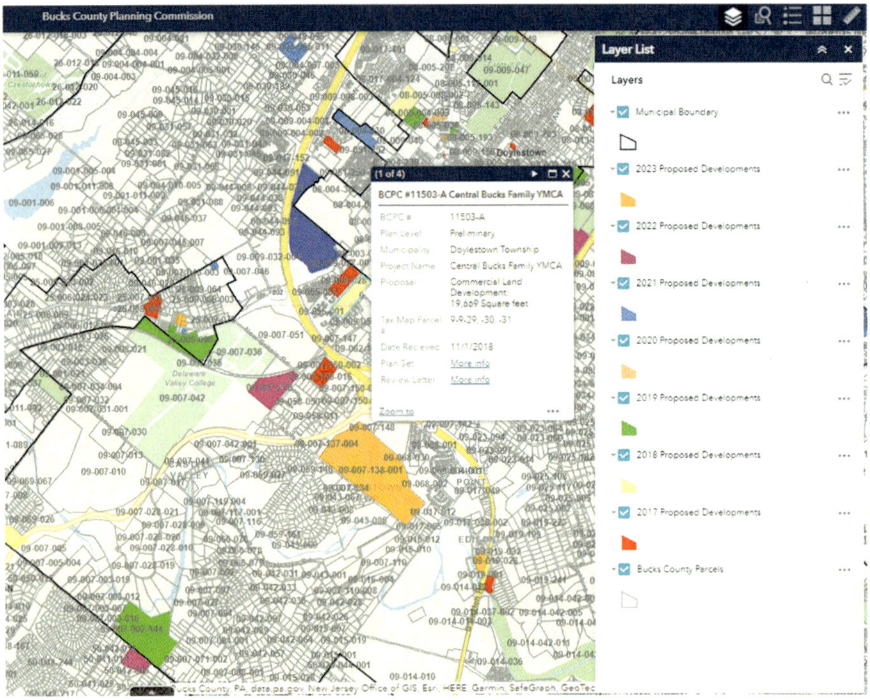


**FIGURE 5.14** Most public visitors to an online GIS will first check out what the GIS has to say about the vicinity of their home

taxpayers while providing a public relations service for the whole community. This is basically a blog like millions of private blogs. But it keeps local citizens in the know and is an easy to link physical with online communities as in the *storymap of the Excelsior neighborhood in San Francisco*, that has been archived by Stanford's University Map Library and may hence be accessible for many years to come.

As discussed in Section 3.5, Esri's storymaps provide a convenient one-stop for creating such effective map-based means of communication. But there are free and open-source alternatives such as *MapStore*.

## 5.5 GIS FOR HOUSING POLICY RESEARCH

In Section 5.3, we introduced basic GIS analysis techniques in the abstract. This section will illustrate the application of these basic techniques with four examples of typical GIS use in housing policy research. Today more than ever, successful public policy depends on high-quality data and the technology that communicates its meaning effectively. Beyond the rational application of scientific or systematic methods, public policy is about values and how values affect, and are affected by, policies. This requires the delivery of credible information in a transparent, understandable form not only to decision makers responsible for adopting policy, but also to various categories of stakeholders whose behavior will be impacted in some way by the policy's implementation.

In order for public policies to be successful, it's important to have good data and technology that can clearly explain what the data means. Public policy isn't just about using science and systematic methods. It's also about values and how those values are impacted by policies. This means that people who make decisions about policies and people who are affected by policies need to have access to reliable and easy-to-understand information.

Anderson (2015) identifies five stages in the policy process:

1. Problem identification
2. Formulation
3. Adoption
4. Implementation
5. Evaluation

Our examples will deal with all of these, but special emphasis will be given to the use of GIS to determine where and when policies are needed, the formulation of public policies, the implementation, and evaluation.

### 5.5.1 Using Cadastral Maps for Problem Identification in Housing Policy Development

Cadastral applications were among the first uses of GIS combining the legal records (attribute data) with the surveying maps – a quintessential example of the georelational principle. Taken by itself, cadasters are little more than repositories with no need for any kind of analysis. These are hyper-local datasets that often are not public because smaller municipalities cannot afford to have their own GIS departments and are using private contractors to develop and maintain a GIS-based cadaster. Increasingly, however, say with the support of their counties, these datasets are being made public and can be used as input for interesting housing-related analyses.

Regardless of provenance, all cadastral datasets have information about the owners, see Figure 5.15 (MapPLUTO, n.d.). Just mapping the top ten landlords makes for interesting insights. Often, these are institutional (governments, churches, and universities) that have an oversized influence on land use planning decisions, but as of late these also include non-traditional landlords such as investment companies.

A second common attribute in a cadastral database is the building age, see Figure 5.16 (MapPLUTO, n.d.). Depending on whether the data has been reconciled with the buildings department (responsible for permitting), this provides valuable information about the nature of the housing stock, from insulation to lead pipes or paint or climate change resiliency.

The number of floors of a building provides useful input to both attempts at neighborhood densification as well as acting as an indicator for the potential for solar roofs (very few buildings with more than five floors have a sloped roof, suitable for the installation of photovoltaic panels; a more thorough analysis would then include aerial imagery, from which one could discern the direction in which a roof slopes, as well as whether it is shaded by trees), see Figure 5.17 (MapPLUTO, n.d.).

In cities that have used GIS for cadastral applications for a while, tax lot change analysis provides valuable insights into the effect of housing policies, see Figure 5.18 (MapPLUTO, n.d.). Information derived from a simple change analysis includes subdivisions, ADUs, zoning changes, etc. The uninitiated would think that all of this can be derived from a spreadsheet as well (basically the attribute component of GIS data) but the crucial information missed by that approach is the determination of "where".

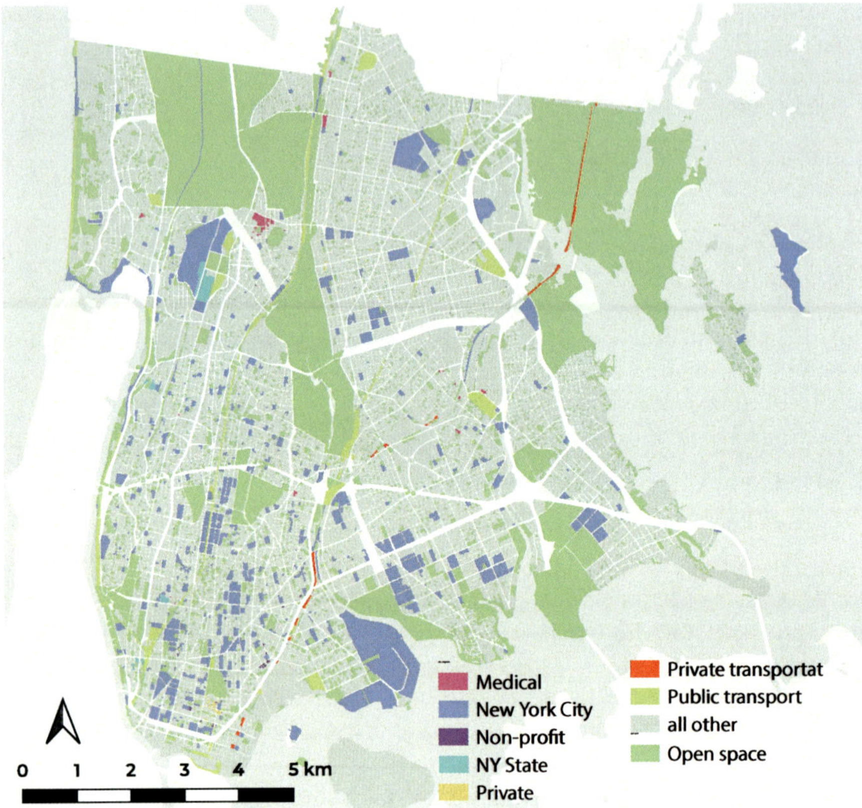


**FIGURE 5.15** Bronx largest property owners

One and the same policy (change) might have very different effects in different parts of a jurisdiction.

Our last example of practical uses of a GIS-based cadaster is the socio-ecological analysis of vacant lots, see Figure 5.19 (MapPLUTO, n.d.). As many municipalities are running out of space for new housing, vacant lots offer at first sight an obvious choice for new developments. But there are always any number of reasons why a lot has not been developed. It may serve as an institutional land bank, it may be in a flood zone, it may be a brown field, or it may just be too small to warrant development without razing buildings on neighboring properties. All of these reasons could be found in a GIS database. It is the linchpin for asking questions beyond the narrow scope of the original creation of the database. This, then, is the argument for establishing such a database in a central IT department which has the capacity to link datasets across functional boundaries.

### 5.5.2 Using GIS to Formulate and Adopt Housing Policy Changes: Gentrification

Understanding displacement is critical given the housing crises around the country: rising rent burdens, homelessness, loss of rent-regulated housing, public housing

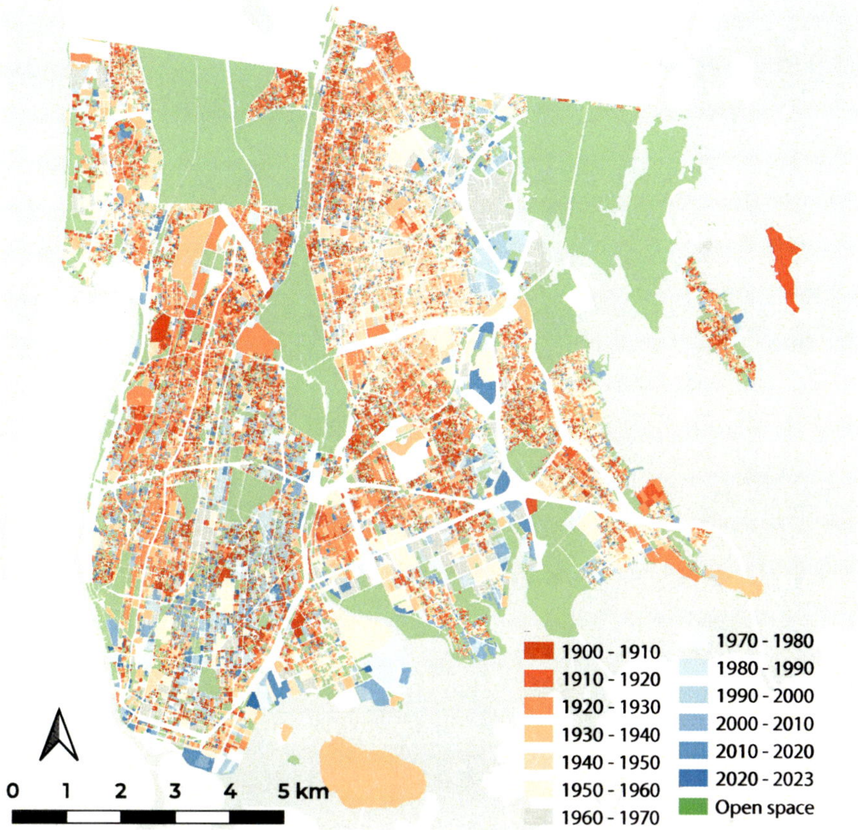


**FIGURE 5.16** Bronx building age

deterioration, and more. We saw in the discussion behind the data for Figure 5.15 that non-for-profit housing initiatives are among the largest property owners in The Bronx, NY. With new federal policies like Opportunity Zones and such local actions that seek to harness market-rate development to boost the supply of affordable housing, it is time to look more carefully at displacement. A popular measure of gentrification is the increase in home values or apartment rents. The problem with that is that property values are almost always going up (everywhere). So, the question then is whether the costs have been going up in a gentrifying neighborhood more than in comparable neighborhoods nearby (with the notion of "nearby" itself being a contentious issue). Slightly more sophisticated is the question of changes in housing affordability (see Chapter 4) and again, its relationship at one location compared to another. At the heart of the gentrification debate, however, is the notion of displacement. The US Census Bureau publishes census tract-level data in response to the question "have you lived in this [area unit] 1/5/10 years ago?" If the answer leans heavily towards shorter time spans, then this may be an indicator for gentrification in a narrower sense. On the other hand, there are numerous neighborhoods around the

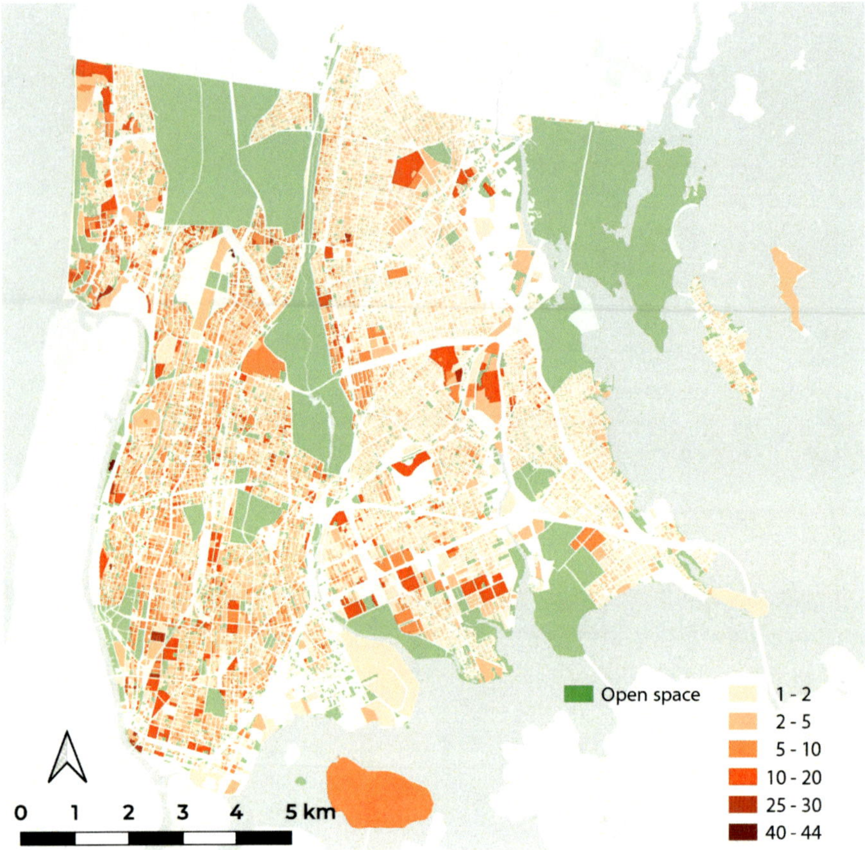


**FIGURE 5.17** Bronx building heights

country (and by the way not limited to urban areas) that have always been transitory, i.e., they serve as landing points for immigrants who then move on after a few years. Even racial or ethnic changes may then be due to international causes and are not suggestive of gentrification.

The following map (Figure 5.20 (US Census, n.d.)) characterized neighborhoods as vulnerable to gentrification if housing sales prices or rent <80% of median, *and* any three of the following four can be observed:

- % low-income households > regional median
- % college educated < regional median
- % renters > regional median
- % nonwhite > regional median

We can then create categories of gentrification by comparing 2000 Census data with 2020 Census data. If a census tract had low-income communities in both years but experiences changes in any of the other bullet points, then this signals ongoing gentrification. If in addition to that, the census tract moved from the low-income to a

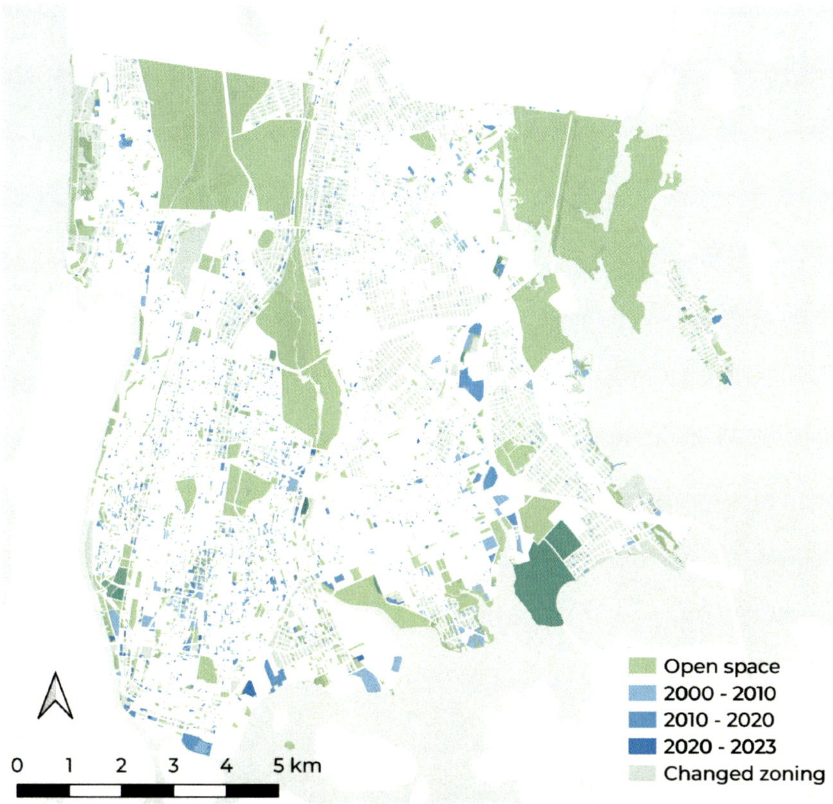


**FIGURE 5.18** Bronx zoning change

middle or high-income class, then this represents an advanced stage of gentrification. "At-risk" are neighborhoods, where the only change so far is above the regional median rise of rents or median property value.

In 2019, some 20% of low-income households or 293,410 people in The Bronx live in low-income neighborhoods at risk of or already experiencing displacement and/or gentrification pressures. We represent as "missing data" those census tracts, where population counts are smaller than 500 residents or the Census Bureau's coefficient of variation suggest a high degree of unreliability.

All of these considerations, however, will only discern the phenomenon after the fact. If gentrification is to be avoided or at least slowed down, then we need to look for indicators of potential future gentrification. A change in amenities (from new green spaces to new transit options (Checker, 2011; Chava and Renne, 2022) may serve as a harbinger of future gentrification. The cumulative effect analysis under the California Environmental Quality Act is a fine example of the utility of having not only a GIS database but, as discussed in Section 3.5.5, also a set of formalized workflows that check for interaction effects of past and present administrative actions (see Figure 5.21 (adapted from Association of Environmental Professionals, 2022)). See the section on GIS challenges in the following chapter for more on geospatial workflow management.

**FIGURE 5.19** Bronx vacant land

### 5.5.3 Using GIS to Evaluate Housing Policy

The term "evaluation" can be applied in a number of different contexts. It may be interpreted as evaluating a situation to understand the severity of a problem, in other words, a needs assessment, or it may be used to evaluate a policy that was established to address the problem. We are going to discuss an example of each in the following pages.

If we are trying to understand the demand for housing in a given area, then we can, following Webster (1993) distinguish between the demand for physical infrastructure and the demand for government regulation such as foreclosure rules. The demand for either may be imputed or based on complaints received. Imputation is based on indicators (see Section 2.2 in Chapter 4) such as overcrowding, heating, plumbing and communication infrastructure, housing affordability, social vulnerability, etc. The result is an inadequate housing map, which may be augmented by point data referencing complaints to a 311 hotline.

Figure 5.22 (San José, 2022; Santa Clara, 2022) shows a mismatch between the imputed and expressed housing demand measures; a discrepancy that is all too common: complaint calls are as much a function of a sense of entitlement or a lack of trust in the efficacy of 311 calls as they are of actual needs. The imputed indicator may hence be better analyzed in light of vulnerable populations such as children, the

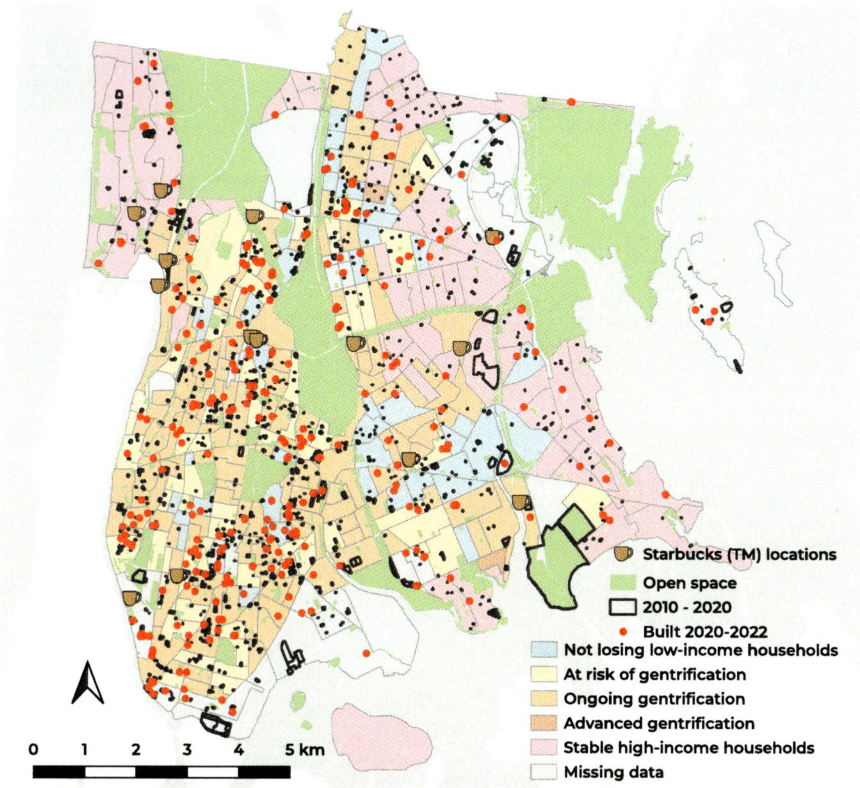


**FIGURE 5.20** Bronx stages of gentrification

elderly, or people with disabilities. (See also Section 6.2 in Chapter 6). The map in Figure 5.22 shows that the majority of severe housing shortages lie in a ring around the city center. How would this change if we weigh the absolute number by accessibility to public transit or the provision of medical services? Even at this basic level of evaluation, there are a multitude of GIS operations to be applied – and none of these questions could be addressed by spreadsheets alone.

GIS-supported housing policy evaluations can be distinguished by time or by space. The former is a classic change analysis of, say, an urban revitalization project, while the latter requires the comparison across a spatial boundary separating the study area into parts where the policy is applied as opposed to those where the policy has not changed (e.g., a transit hub on the edge of a municipal jurisdiction). For an evaluation along a temporal axis, the process is similar to the identification of milestones and deliverables in project management. At each stage of the project, inventories are taken and then compared.

Jurisdictional boundaries lend themselves to the planning equivalent of working with control groups in a medical experiment. Many metropolitan areas in the United States have beltways that separate a larger city from its surrounding municipalities. As public transit follows these existing corridors and transit-oriented development fosters densification around transit stations (see Section 3.2.6), these become living laboratories for the effect of different housing policies as they are implemented by

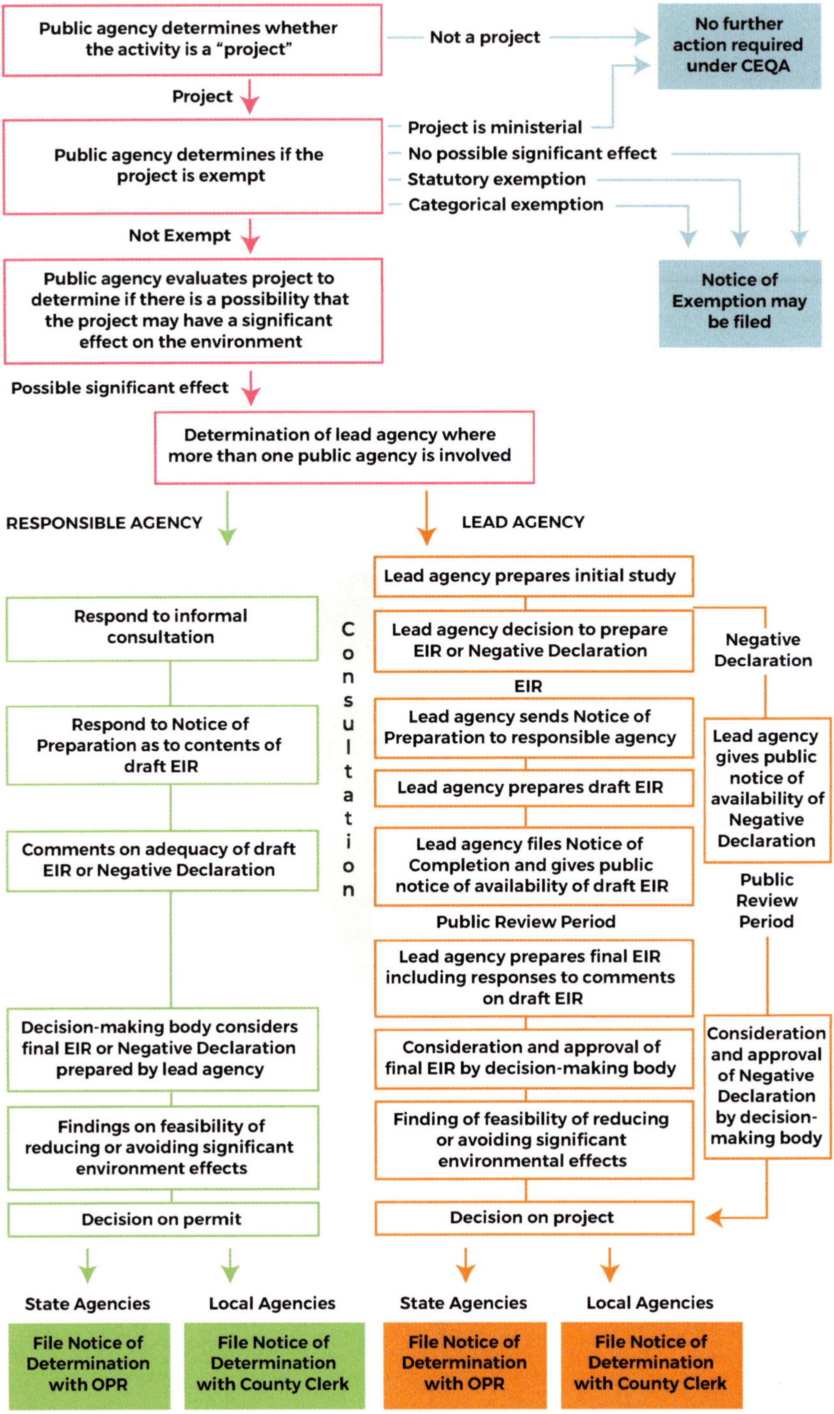


FIGURE 5.21 The CEQA flowchart

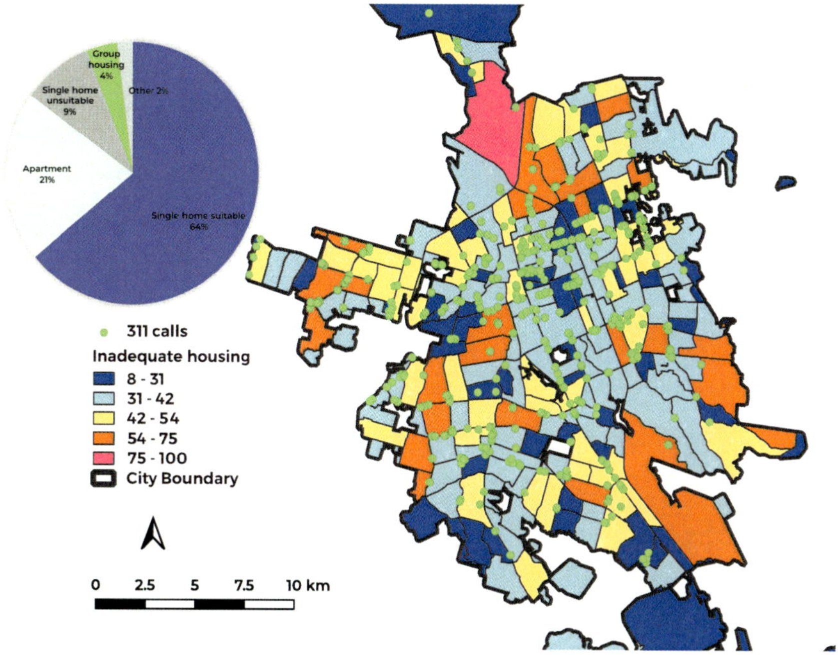


**FIGURE 5.22** Tracking community concerns: map and pie chart revealing 311 calls and inadequate housing issues in San Jose, CA

varying local authorities in their vicinity. Auerbach et al. (2020), for instance, report on the use of GIS to compare the effect of an anti-displacement tax fund on West-Atlanta neighborhoods that participate in the effort compared to those who do not.

## 5.6 ADVANCED TECHNIQUES

The previous two sections provided a pretty thorough introduction to GIS for housing policy research. We laid the technical foundations in Chapter 4 and then delved into the necessary concepts of GIS data models and the main (most commonly used) analysis operations. These sections, in conjunction with a bit of trial and error or learning by doing, will enable diligent readers to use GIS in their everyday housing policy work. The remainder of this chapter is a high-level overview of more advanced GIS techniques available to seasoned housing researchers. This section covers material commonly taught in one or two graduate-level GIS courses but can, of course, not be as thorough. Novices are invited to read this section to learn about topics that may relate to experiences outside the geospatial realm. Readers with some GIS experience will discover applications that go beyond the traditional buffer and overlay paradigm. This section is heavily annotated with links for further readings.

### 5.6.1 Dasymetric Mapping and Pycnophylactic Interpolation

The term dasymetric mapping (DM) is misleading as it suggests a visualization technique. While it can be used as such, its importance lies mainly in the impact it has

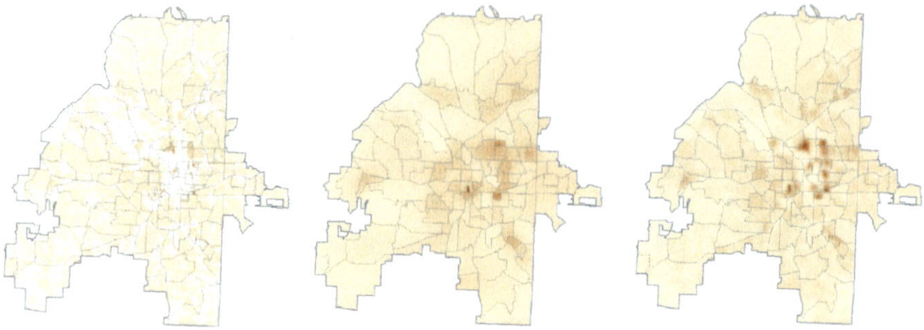

FIGURE 5.23 Atlanta population (a) mapped dasymetrically (b) interpolated pycnophylactically, and (c) with both techniques combined. Based on Kim and Yao (2010)

on analyses. DM is essentially a response to the ecological fallacy of trying to reason about something specific based on only general data. Take US Census tract data for instance. As with all polygon data, it says nothing about how the phenomenon is distributed within a census tract. But we do know that the population the data is based on are residences. And we know that (with very few exceptions) people do not live on water or in parks, parking lots, etc. So, we can redistribute our population-based Census data to those parts of a tract that remain after we have subtracted (another basic GIS operation) the uninhabitable areas. We may even, if we have access to building footprint data, limit the distributions to the building footprints themselves. The result is a much more realistic representation that deals with one aspect of the modifiable area unit problem (MAUP).

The other aspect is that of arbitrarily drawn boundaries that artificially separate continuous phenomena. Returning to our census population example, it is just not reasonable to assume that population characteristics change at the boundary between two tracts. Tobler (1979) developed a technique called *pycnophylactic interpolation* (PI) that takes information about the distribution of a phenomenon in neighboring regions to redistribute the data within each region (e.g., a tract) and create smooth transition across boundaries. The implementation requires translating vector to raster data and having local knowledge about the existence of discontinuities such as rivers, parks, railway lines, etc., all of which would render any interpolation assumption incorrect.

Kim and Yao (2010) present examples that combine dasymetric mapping (DM) and pycnophylactic interpolation (PI) to create data that seems to mysteriously be of much higher accuracy than any input data, see Figure 5.23 (US Census, 2010; Atlanta Regional Commission, 2010). This is reminiscent of the Bayesian approach, where the incorporation of auxiliary data (such as land use) results in much improved interpolation results. It is particularly useful in situations where we try to work with relatively coarse data like from the public health sector. Rather than trying to take our analysis to the parcel level, we can try to improve those coarser datasets so as to not water down our results. In addition to the need to handle the transition between raster and vector data and to find software that performs the pycnophylactic interpolation (R or Python), the main concern is that the processes and results of either DM or PI are consistent with the conceptual model of the researcher. This means that she has to be aware of the assumptions that underlie the creation of the original datasets, in particular its spatial support.[4]

### 5.6.2 Pattern and Cluster Analysis

We have, by now, presented dozens of maps to illustrate one argument or the other. The built in assumption has been that the map shows the distribution of a particular phenomenon and that the patterns on the map are (i) real, i.e., they can be observed if we visit the place depicted and (ii) are pertinent or (statistically) relevant. The former is difficult to maintain because most of our maps are actually abstractions that have to be translated back into the experiential knowledge of a local observer. The latter takes us into the realm of spatial statistics, which is necessitated by the fact that humans have an uncanny ability to detect patterns where objectively there aren't any (Goldstone and Barsalou 1998; Reber et al. 1998; Rensinck and Baldridge 1998). In other words, we are neurologically hardwired to detect patterns because they are the basis of object recognition and hence our ability to navigate and make sense of the world. This is then, where pattern and cluster analysis come to bear.

The majority of applications are based on point data (e.g., crime locations, 311 calls, grocery stores) because the geometries are easier to run calculations on than with linear or areal features. And here, it is easier dealing just with locations rather than weighing them by some attribute value (e.g., square footage of the grocery store). The question of spatial support raises its head again because something as innocuous as bus stops cannot be randomly distributed as they are spatially constrained by the road network. It is the lack of randomness in urban spatial phenomena that invites spatial statistical analysis. All spatial pattern analyses are about comparing the observed pattern to a set of random patterns to then determine whether the observed one is likely to be random or not. If it could be random (without some chosen confidence interval) then we declare the pattern to not be statistically significant. Matters are complicated by (i) the definition of the boundary of our study area (for instance, we don't expect burglaries to occur inside lakes or water bodies, although theft of fish or water would be another matter entirely) and (ii) the scale of analysis. Something may look like a pattern at one scale but not at another. This, however, points to one of the purposes of the analysis in the first place. Just identifying a pattern is hardly enough; we then want to determine what are the drivers behind the distributions that we observe – and scale dependency helps us to limit the range of possible drivers.

When we determine that our observed pattern is not consistent with randomness, there are two possibilities: the observed pattern may exhibit signs of (i) clustering or of (ii) dispersion. Small amounts of either are normal and would be expected in a random distribution but consistent or strong patterns of clustering or dispersion (e.g., the distribution of black and white fields on a chessboard) point to some forcing factor.

A cluster is described as the intensity of the phenomenon: the more observations in a small area, the more intense the phenomenon (crime, Covid-SARS cases, etc.). This is measured by a so-called kernel density function, where a small (size to be determined and usually the procedure is repeated for many different sizes) search window is continuously moved over the study area to count the number of observations within the search window. The systematic application of varying search window sizes helps with the determination of the pertinent scale of the observed clustering.

The detection of patterns in areal data (e.g., census tracts) requires a discussion of spatial autocorrelation. The same Tobler of pycnophylactic interpolation was coined in an obscure article in 1970 *The First Law of Geography*, which states "everything is related to everything else, but near things are more related than distant things". It underlies all work in spatial analysis and is the basis for any scientific approach to geography (including GIS) because without it phenomena would be distributed randomly in space and we would have no way to systematically reason about them. Statistically, the first law captures autocorrelation, i.e., the correlation of a variable with itself as a function of distance.[5] The analogy of a chessboard helps again. The black and white fields are perfectly negatively autocorrelated, i.e., every white field shares on all sides boundaries with black fields and vice versa. The position of the figures at the beginning of the game is exactly the opposite: all white figures have only white neighbors and all black figures on black ones. This simple arrangement is harder to discern when the areas are irregular (like Census area units). We then have to establish who is a neighbor of whom (the topological relationships we discussed in Section 5.3 of this chapter), which is encoded in the form of weight matrices that establish the degree of neighborship. There are multiple measures of spatial autocorrelation with the most common one probably being Moran's *I*, which is a global measure of the relationship between spatial proximity and variable similarity. A local version known as local indicator of spatial association (LISA) captures the difference between the spatial autocorrelation of a small set of neighbors compared with the global measure. It is used to identify so-called hotspots and coldspots (see Figure 5.24 (San José Bikeways, 2022).

### 5.6.3 Geographically Weighted Regression

> Imagine reading a book on the climate of the United States which contained only data averaged across the whole country, such as mean annual rainfall, mean annual number of hours of sunshine, and so forth. Many would feel rather short-changed with such a lack of detail. We would suspect, quite rightly, that there is a great richness in the underlying data on which these averages have been calculated; we would probably want to see these data, preferably drawn on maps, in order to appreciate the spatial variations in climate that are hidden in the reported averages. Indeed, the averages we have been presented with may be practically useless in telling us anything about climate in any particular part of the United States. It is known, for instance, that parts of the north-western United States receive a great deal more precipitation than parts of the Southwest and that Florida receives more hours of sunshine in a year than New York. In fact, it might be the case that not a single weather station in the country has the characteristics depicted by the mean climatic statistics.

This is the introductory paragraph for *Geographically Weighted Regression* by Fotheringham et al. (2002). And the paragraph describes succinctly one of the main points that we are trying to make in this volume, namely that (i) space/location matters, (ii) that things are not uniformly distributed throughout a region, and (iii) that we have to distinguish between local and global phenomena, where the definition of what constitutes local is variable. This then begs the question how to define a *local* regime or realm of influence. This is exactly what geographically weighted regression (GWR) is good for to answer.

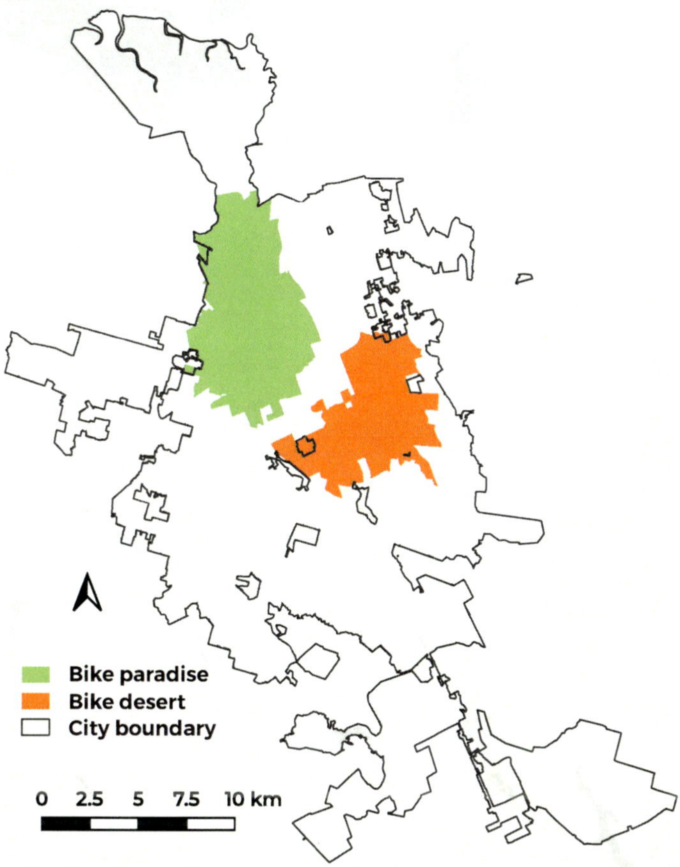


**FIGURE 5.24** San Jose bike desert. From Zandiatashbar et al. (2023)

To appreciate the problem that the GWR is trying to solve, let's have a look at a regression model that tries to explain house prices based on a few explanatory variables such as size of the property, amenities, building age, and unemployment rate. A traditional regression model would give us an equation like

$$p = \alpha_0 + \alpha_1 \text{ propsize} + \alpha_2 \text{ amenities} + \alpha_3 \text{ bldage} + \alpha_4 \text{ unemploy} + \varepsilon$$

The error term $\varepsilon$, covering the unexplained component(s) of our model, would then be assumed to be randomly distributed over our study area. As it turns out, however, this is not the case, and it is easily visualized by mapping the difference between the expected and the observed values as in Figure 5.25 (US Census, n.d.; NYC Transit, 2020).

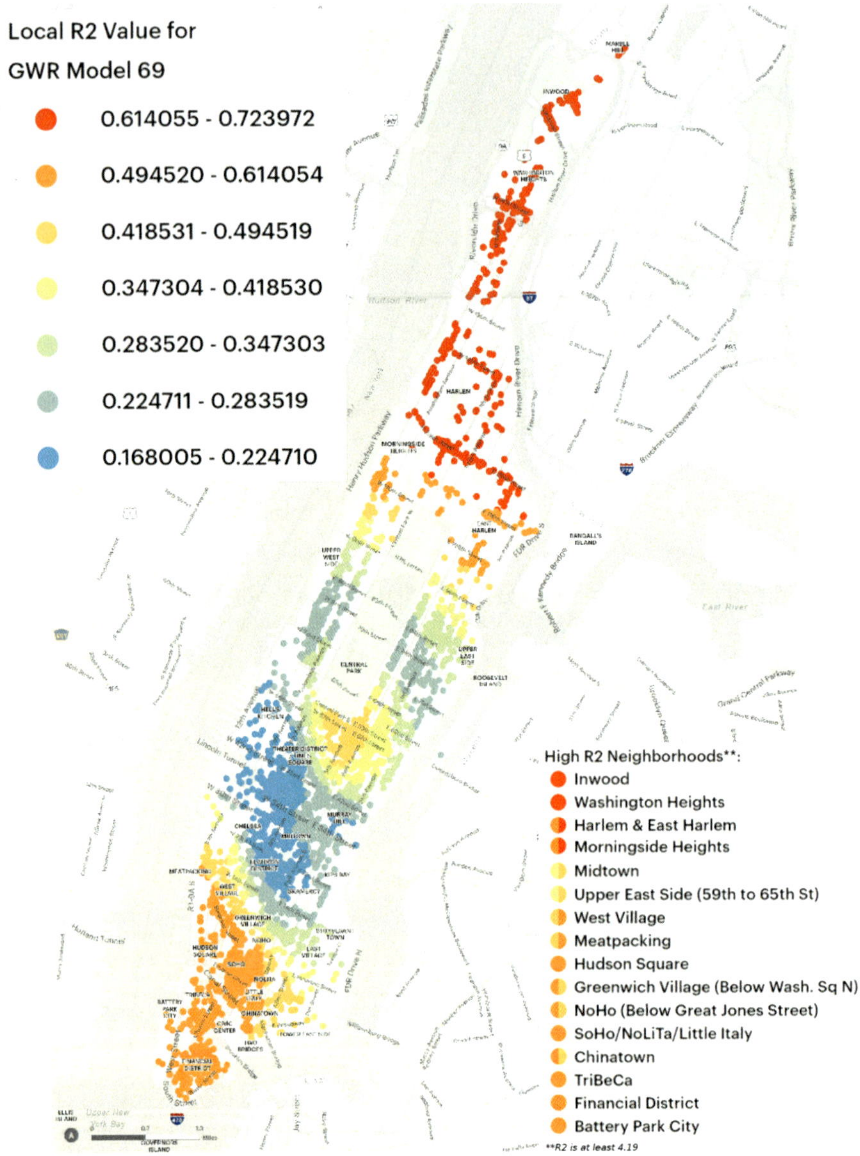


**FIGURE 5.25** Non-random (spatially auto-correlated) distribution of residuals in a global regression model

Clearly, when we look at Figure 5.25, we can detect that the residuals are not randomly distributed as we would expect from a random process. We could verify this impression by performing a Moran's *I* spatial autocorrelation test. In other words, the contribution of individual explanatory variables varies over our study area, e.g., the effect of property size on the final price is different in one part of the study area compared to some other. An observant reader might object that this may be due to the MAUP, and if we had chosen area boundaries appropriately, then the map would look very different. However, this is not the case as can be shown if we do not work with polygons but with point data (each individual home sale), which would result in a density map of residuals.

The next logical step would then be to create individual local regression models for each of the ZIP code areas in Figure 5.18. In addition to this getting rather tedious, we would now indeed run into the MAUP, so this is not a practical solution – especially if the footprints for the explanatory variables are varied. The solution comes in the form of a technique adopted from point pattern analysis called moving window regression. A search window of a fraction of the size of the study area is continuously moved over the study and the regression is applied to all the observations that fall within the search window. The MAUP is then resolved by not having the search window jump by the width of its size but say by 1/10th of its size. This smoothes the differences between the regression results and does not assume any boundaries. This is computationally intensive and we would leave it at that if we have a good idea of how far neighborhood effects extend for a particular variable. If this is not the case, then we would run the same GWR procedure with varying window sizes and instead of square windows would employ so-called kernels with varying distance decay functions (Figure 5.26).

The effect of this procedure is three different outcomes, two of which are important, while the third one is contentious. First, when we now map the residuals, we will find that there is no spatial autocorrelation to them and that they are indeed randomly distributed – as we should expect from a regression model. Second, the GWR gives us areas of likewise spatial regimes where the respective regression equations are either the same or very similar. These areas are not the result of any boundaries in the input data but constitute a regionalization of our dependent variable. The importance of this statement is hard to overemphasize; the GWR tells us where, in spite of the curse of spatial variation, we can expect uniform behavior in response to our

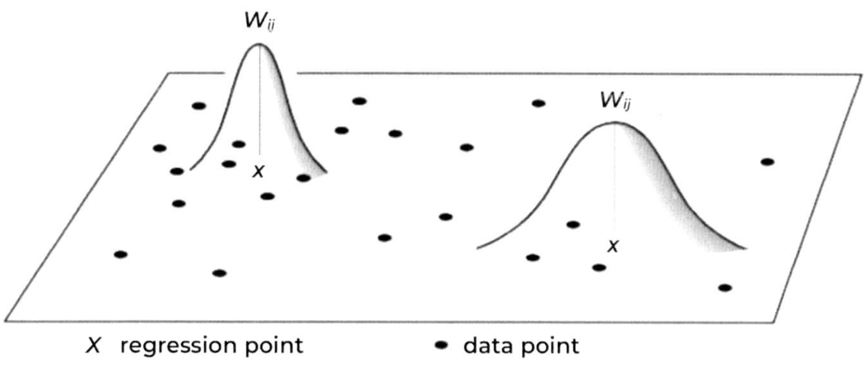


**FIGURE 5.26** Varying kernel sizes to emphasize the contribution of neighboring observations as a function of distance

policy decisions. Finally, and this is the contentious part, the GWR gives us usually much improved $r^2$ values that make us feel good but that many in the community of professional statisticians declare to be unfounded. If the purpose of our analysis is a sound explanatory model, then we have to resort to the spatial regression techniques of the following sub-section. But the value of outcome (2), the regionalization of our research question should not be underestimated.

### 5.6.4 Spatial Regression

In the most general sense possible, a regression equation describes the relationship between a dependent variable whose value we want to predict on the left-hand side and any number of independent variables that serve to explain the outcome as in this equation:

$$\text{outcome} = \alpha_1 var_1 + \alpha_2 var_2 + \alpha_3 var_n + \varepsilon$$

In non-spatial applications, the parameters $\alpha$ provide a kind of weight (which may also be negative as when higher incomes usually suggest fewer single parents).[6] It is good statistical practice to work with variable values that have been transformed to standardized ranges to ascertain that the parameters relate appropriately to each other. The additional twist in spatial versions of a regression equation is that each $\alpha$ is in turn adjusted by what is known as a spatial weight matrix. The spatial weight matrix is a construction that specifies the influence that the value of one observation has on its neighbors and is usually distance-weighted, i.e., observations further away have a lesser influence (Tobler's First Law). There are a multitude of methods to create such a spatial weight matrix, depending on the type of geometry as well as how many neighbors should be incorporated and the reader is referred to standard textbooks such as LeSage and Pace (2009) and Anselin and Rey (2010), or Anselin and Rey (2014).

The obvious reason for the construction of the spatial weight matrix is to deal with spatial autocorrelation; something that is seen as a nuisance in traditional statistics but is now employed as an additional piece of information. The GWR from the previous subsection implicitly creates an optimized spatial weight matrix but does not export it for further exploration or comparison. In another twist, spatial influences may not just impact the values of each explanatory variable but may also be hidden in the error term $\varepsilon$. Models addressing the former are referred to as spatial lag models (explaining the influence that neighbors have), while the latter is known as spatial error models. The spatial lag $y_{lag\text{-}i}$ is

$$y_{lag-i} = \sum_j w_{ij} y_j$$

where $y_{lag\text{-}i}$ is the spatial lag of variable $y$ at location $i$, and $j$ sums over the entire dataset. For spatial error models the traditional e is replaced with $u_{lag\text{-}i} + \varepsilon_i$.

Traditional GIS are not made for this but many of the bigger statistics programs have modules for spatial regression; none more so than the statistics package R.

## 5.7 FURTHER READING

These last two chapters concentrated on the technical aspects of GIS for housing policy research. Readers who want to go beyond what has been presented here will want to peruse some of the readings suggested in the next free paragraphs. However, before you take off to another book, let's have a look at what the next chapter has to offer.

While Chapters 1–3 provided an overview of the housing policy landscape and the kinds of problems we are trying to solve, Chapter 4 introduced us to the geospatial data that then allows us to make use of the unique capabilities of GIS in Chapter 5. One of the tenets of this book is that the geographic perspective of spatial differentiation has been underdeveloped in much of the housing policy literature. Many problems can only be addressed if they are seen both in concert with the perspectives of related fields as well as the unique set of circumstances/conditions that makes each location unique. With this in mind, we can now apply the GIS tools introduced here to the big challenges that every housing researcher is confronted with in the 21st century. Regardless of whether we want to overcome the single-family residential paradigm, modernize housing and neighborhood design, deal with the changes of mobility patterns brought about by the diversification and hybridization of work, combat homelessness and housing insecurity, deal with climate change, public health or public safety, GIS lies at the center of each solution space. In Chapter 6, we will illustrate through numerous examples how GIS is used to address each of these challenges.

### 5.7.1 GIS Models

A good overview of vector data formats can be found in Diamond (2019), while the corresponding article for raster formats is Williams (2019). Conceptual data models, including tools and languages to compile them, are well covered in Nyerges (2017a). From a GIS project development perspective, this should precede the choice of logical data model described by the same author in (2017b).

A very brief introduction to conceptual ways of organizing spatial data is Varanka's (2021) article, however, the reader might want to skip right down to the end of this encyclopedia entry to find truly further readings; it lists many classics that should be on the shelf of every GIS practitioner. Two specific data models discussed in our volume are the raster and the vector model. A nice overview of the former is Pingel (2018), which is complemented by Albrecht's (2022) discussion of entity-based models. Albrecht's article also makes for a good entry point to the next section on basic GIS analysis operations.

### 5.7.2 Basic GIS Analysis Operations

Spatial neighborhoods can be defined in many different ways and Mu and Holloway (2019) provide a nice overview. Interestingly, they miss a crucial body of work epitomized by the Laval school of geomatics. Gold's (2016) article on tessellations would be a good representative of that line of thinking. Another fundamental approach to understanding basic GIS analyses is set theory. Arlinghaus' (2019) article is a good starting point. This leads directly to overlay analysis as introduced by Cai (2022), the counterpart to which would be Li's (2017) entry on buffering.

### 5.7.3 Advanced GIS Techniques

A good introduction to dasymetric mapping is Mennis' (2017) encyclopedia entry. It builds on Tobler's (1979) article on pycnophylactic interpolation, which is eminently readable in spite of its publication in the Journal of the American Statistical Association. Pattern and spatial cluster analysis are common techniques in landscape ecology and crime analysis. There are thousands of applications but the original description in McGarigal and Marks (1995) remains the go-to reading on this subject.

An excellent user-friendly introduction to a range of spatial (statistical) analysis techniques is the GeoDa software developed by the Center for Spatial Data Science at the University of Chicago. GeoDa incorporates a range of spatial analysis methods in a very user friendly way, one of which is Local Indicators of Spatial Association, first described by Anselin (1995). One technique not covered by GeoDa is Geographically Weighted Regression (GWR), epitomized by Fotheringham et al. (2003). Although eminently readable, readers of this volume might want to start with Sachdeva and Fotheringham's (2020) overview. Chakraborty and McMillan's (2022) article entitled "Is Housing Diversity Good for Community Stability?" is a nice example of the application of *spatial regression* in housing research.

## NOTES

1. The same problem occurs in the world of mankind as well; see, for example, the ill-defined boundaries of neighborhoods or regions such as the boundary between the eastern United States and the Midwest.
2. The American meteorologist Alfred Thiessen (1911) and the Ukrainian mathematician Georgy Voronoi (1908) introduced these structures to a geophysical community at roughly the same time without knowing about the respective other's work. They were both preceded by the German mathematician Dirichlet, who in 1850 in his *Über die Reduction der positiven quadratischen Formen mit drei unbestimmten ganzen Zahlen* defined what in mathematics is known as Dirichlet regions.
3. Database aficionados would beg to differ as all of this can also be done with spatial SQL.
4. In mathematics, the support of a real-valued function $f$ is the subset of the domain containing the elements which are not mapped to zero. If the domain of $f$ is a topological space, the support of $f$ is instead defined as the smallest closed set containing all points not mapped to zero.
5. Outside of geospatial applications, auto-correlation is typically understood to be the correlation of a variable with itself as a function of a lag or distance in time.
6. A negative variable weight $\alpha_n$ indicates that the outcome increases as the variable value decreases. If, for example, the outcome variable is median area income, then a smaller number of single parents typically results in a higher area income (and vice versa).

## REFERENCES


Albrecht, J, 2022. "Entity-Based Models". In: John P. Wilson (Ed.), *Geographic Information Science & Technology Body of Knowledge*. doi:10.22224/gistbok/2022.2.11.

Anderson, J, 2015. *Public Policymaking*. Stamford, CT: Cengage Learning.

Anselin, L, 1995. "Local Indicators of Spatial Association-LISA". *Geographical Analysis*, 27: 93–115, doi:10.1111/j.1538-4632.1995.tb00338.x.

Anselin, L, and Rey, S, 2010. *Perspectives on Spatial Data Analysis*. Berlin/Heidelberg: Springer. doi:10.1007/978-3-642-01976-0.

Anselin, L, and Rey, S, 2014. *Modern Spatial Econometrics in Practice: A Guide to GeoDa, GeoDaSpace and PySAL*. Chicago, IL: GeoDa Press LLC.

Arlinghaus, S, 2019. "Set Theory". In: Wilson, J (Ed.), In: *The Geographic Information Science & Technology Body of Knowledge* (2nd quarter, 2019 edition). doi:10.22224/gistbok/2019.2.1.

Association of Environmental Professionals, 2022. *2022 CEQA California Environmental Quality Act Statute and Guidelines*. Palm Desert: California AEP.

Baltimore, DHCD. *CoDe Map*, n.d. https://cels.baltimorehousing.org/codemapv2ext/, 14 July 2023.

Bucks County Planning Commission, n.d. *Proposed Subdivisions and Land Developments*. https://bucksgis.maps.arcgis.com/apps/webappviewer/index.html?id=f58e99f72c4241ebbe309e08d6e42198, 14 July 2023.

Cai, H, 2022. "Overlay". In: Wilson, J (Ed.), *The Geographic Information Science & Technology Body of Knowledge* (1st quarter 2022 edition). doi:10.22224/gistbok/2022.1.2.

Chava, J, and Renne, J, 2022. "Transit-Induced Gentrification or Vice Versa?" *Journal of the American Planning Association*, 88(1): 44–54.

Checker, M, 2011. "Wiped Out by the 'Greenwave': Environmental Gentrification and the Paradoxical Politics of Urban Sustainability". *City & Society*, 23(2): 210–29. doi:10.1111/j.1548-744X.2011.01063.x

Chu, M, Fenelon, A, Rodriguez, J, Zota, and Adamkiewicz, G, 2022. "Development of a Multidimensional Housing and Environmental Quality Index (HEQI): Application to the American Housing Survey". *Environmental Health*, 21:56. doi:10.1186/s12940-022-00866-8.

Chester County Planning Commission, n.d. *Chester County Planning Commission*. https://www.chescoplanning.org/planreview/Maps.cfm, 14 July 2023.

Devillers, R and Jeansoulin, R, 2006. "Fundamentals of Spatial Data Quality". London: Wiley. doi:10.1002/9780470612156.

Diamond, L, 2019. "Vector Formats and Sources". In: Wilson, J (Ed.), *The Geographic Information Science & Technology Body of Knowledge* (4th quarter, 2019 edition), https://doi/10.22224/gistbok/2019.4.8.

Donnelly, F, 2022. *US Census Data: Concepts and Applications for Supporting Research*. American Library Association Library Technology Reports (vol. 58, no. 4). Chicago, IL: ALA TechSource.

FEMA, 2022. "Hazus 6.0 Baseline Data Updates". *FEMA Factsheet*. https://www.fema.gov/sites/default/files/documents/fema_hazus-6-data-updates-factsheet.pdf, last accessed 4 December 2022.

Fotheringham, A, Brunsdon, C, and Charlton, M, 2003. *Geographically Weighted Regression: The Analysis of Spatially Varying Relationships*. New York: John Wiley & Sons.

Geo, DA, 2023. https://geodacenter.github.io/, last accessed 23 May 2023.

Gold, C, 2016. "Tessellations in GIS: Part I-putting it all together". *Geo-Spatial Information Science*, 19(1): 9–25, doi:10.1080/10095020.2016.1146440

Goldstone, R, and Barsalou, L, 1998. "Reuniting Perception and Cognition". *Cognition*, 65: 231–262.

Goodchild, M, Li, W, and Tong, D, 2022. "Introduction to the Special Issue on Scale and Spatial Analytics". *Journal of Geographical Systems*, 24: 285–289. doi:10.1007/s10109-022-00391-9

Guptill, S, and Morrison, J, 1995. *Elements of Spatial Data Quality*. Amsterdam:Elsevier.

Hirschman, C, Alba, R, and Farley, R, 2000. "The Meaning and Measurement of Race in the U.S. Census: Glimpses into the Future". *Demography*, 37: 381–393.

HUD PD&R Edge, 2017. *Defining Housing Affordability*. PD&R Edge. Washington, DC: U.S. Department of Housing and Urban Development's (HUD's) Office of Policy Development and Research (PD&R). https://www.huduser.gov/portal/pdredge/pdr-edge-featd-article-081417.html, last accessed 04 March 2023.

Jacobs, K, and Mitchell, S, 2020. "OpenStreetMap Quality Assessment Using Unsupervised Machine Learning Methods". *Transactions in GIS*, 24: 1280–1298. doi:10.1111/tgis.12680.

LeSage, J, and Pace, R, 2009. "Introduction to Spatial Econometrics". New York: Chapman and Hall/CRC. doi:10.1201/9781420064254.

Li, X, 2017. "Buffers". In: Wilson, J (Ed.), *The Geographic Information Science & Technology Body of Knowledge* (4th quarter, 2017 edition). doi:10.22224/gistbok/2017.4.10.

MapPLUTO, n.d. *New York City Department of Urban Planning and Development parcel-level database*. Online resource available at https://www.nyc.gov/site/planning/data-maps/open-data/dwn-pluto-mappluto.

McGarigal, K, and Marks, B, 1995. *FRAGSTATS: Spatial Pattern Analysis Program for Quantifying Landscape Structure*. USDA Forest Service General Technical Report PNW-351, Corvallis. https://www.fs.usda.gov/pnw/pubs/pnw_gtr351.pdf, last accessed 22 December 2022.

Mennis, J, 2017. "Dasymetric Mapping". In: Richardson, D, Castree, N, Goodchild, M, Kobayashi, A, Liu, W, and Marston, R (Eds.), *International Encyclopedia of Geography: People, the Earth, Environment and Technology*. Hoboken, NJ: Wiley. doi:10.1002/9781118786352.wbieg0443.

Missouri Census Data Center, 2022. "Intro to Census Geography, Summary Levels, and GeoIDs". https://mcdc.missouri.edu/geography/sumlevs/, last accessed 4 December 2022.

Mu, L, and Holloway, S, 2019. "Neighborhoods". In: Wilson, J (Ed.), *The Geographic Information Science and Technology Body of Knowledge* (1st quarter, 2019 edition). doi:10.22224/gistbok/2019.1.11.

National Institute of Standards and Technology (NIST), 2021. *Compliance FAQs: Federal Information Processing Standards (FIPS)*. https://www.nist.gov/standardsgov/compliance-faqs-federal-information-processing-standards-fips, last accessed 4 December 2022.

NYC Open Data, n.d. NYC Open Datamine, an online resource available at https://opendata.cityofnewyork.us/data/.

NYC Transit, 2020. *Turnstyle passenger counts at New York City subway stations*. Online resource available at https://qri.cloud/nyc-transit-data/turnstile_daily_counts_2020.

Nyerges, T, 2017a. "Logical Data Models". In: Wilson, J (Ed.), *The Geographic Information Science & Technology Body of Knowledge* (1st quarter, 2017 edition), doi:10.22224/gistbok/2017.1.2.

Nyerges, T, 2017b. "Conceptual Data Models". In: Wilson, J (Ed.), *The Geographic Information Science & Technology Body of Knowledge* (1st quarter, 2017 edition). doi:10.22224/gistbok/2017.1.3.

OGC, 2023. *E-Learning Documentation*. https://opengeospatial.github.io/e-learning/geopackage/text/contents.html, last accessed 24 May 2023.

Openshaw, S, and Taylor, P, 1979. "A Million or So Correlation Coefficients: Three Experiments on the Modifiable Areal Unit Problem". In: Wrigley, N (Ed.), *Statistical Methods in the Spatial Sciences*, pp. 127–144. London: Pion.

OPR, 2023. *CEQA: The California Environmental Quality Act*. Sacramento, CA: Governor's Office of Planning and Research. https://opr.ca.gov/ceqa/, last accessed 30 May 2023.

Pingel, T, 2018. "The Raster Data Model". In: Wilson, J (Ed.), *The Geographic Information Science & Technology Body of Knowledge* (3rd quarter, 2018 edition). doi:10.22224/gistbok/2018.3.11.

R Spatial, 2023. https://cran.r-project.org/web/views/Spatial.html, last accessed 30 May 2023.

Reber, R, Schwarz, N, and Winkielman, P, 1998. "Effects of Processing Fluency on Affective Judgments". *Psychological Science*, 9: 45–48.

Rensink, R, and Baldridge, G, 2010. "The Perception of Correlation in Scatterplots", *Computer Graphics Forum*, 29(10): 1203–1210.

Sachdeva, M, and Fotheringham, A, 2020. "The Geographically Weighted Regression Framework". In: Wilson, J. (Ed.), *The Geographic Information Science and Technology Body of Knowledge* (4th quarter, 2020 edition), doi:10.22224/gistbok/2020.4.7.

San José, 2022. *Database of 311 calls to the San Jose, CA, call center*. Online resource available at https://311.sanjoseca.gov/.

San José Bikeways, 2022. *Bicycle network dataset*. Online resource available at https://data.sanjoseca.gov/dataset/bikeways.

Santa Clara, 2022. *Database of housing quality standards inspections conducted by the Housing Authority of Santa Clara County*, available at schousingauthority.org.

Sparx Systems, 2022. *Guide to Business Modeling*. https://sparxsystems.com/resources/user-guides/16.0/guidebooks/business-modeling-techniques.pdf, last accessed 12/04/2022.

Tobler, W, 1979. "Smooth Pycnophylactic Interpolation for Geographical Regions". *Journal of the American Statistical Association*, 74(367): 519–530. doi:10.2307/2286968

US Census, n.d. *Generic US Census data*. Online resource available at https://data,census.gov.

US Census, 2010. *Block-level total population 2010*. Online resource available at https://data.census.gov.

Varanka, D, 2021. "Data Properties". In: Wilson, J (Ed.), *The Geographic Information Science & Technology Body of Knowledge* (1st quarter, 2021 edition). doi:10.22224/gistbok/2021.1.15.

Westlaw, 2023. *Guidelines of the CEQA, in the California Code of Regulations, Title 14, Natural Resources*. Division 6 Resources Agency. https://govt.westlaw.com/calregs/Index?transitionType=Default&contextData=%28sc.Default%29, last accessed 7 May 2023.

Williams, C, 2019. "Raster Formats and Sources". In: Wilson, J (Ed.), *The Geographic Information Science & Technology Body of Knowledge* (4th quarter, 2019 edition). doi:10.22224/gistbok/2019.4.11

Zandiatashbar, A, Albrecht, J, and Nixon, H, 2023. *A Bike System for All in Silicon Valley: Equity Assessment of Bike Infrastructure in San José, CA*. San José, CA: San José State University, College of Business, Mineta Transportation Institute.

# 6 Directions for Future Research

## 6.1 TRANSFORMING THE SINGLE-FAMILY RESIDENTIAL UNIT

Single-family residential units make up the largest percentage of residential housing units in the United States. Thus, redesigning private homes to make them adaptable for life in 2050 and beyond requires a concerted and collaborative effort in which housing policy specialists and affordable housing advocates must consciously work with urban planners, architects, landscape architects, interior designers, engineers, and the building trades.

Architects, interior designers, and home builders often use the claim that they deliver *what the market wants*. One of these adaptations relates to the size of a single-family unit. According to data analyzed by the National Association of Home Builders, the average single-family unit was about 2,200 square feet in 1999 and that number has gradually increased to about 2,500 square feet in 2022. The recent pandemic experiences of 2020–2022 have prompted homeowners to seek out more open spaces and natural light/ventilation within and around their residential units, expanding the footprint of the house. Privileged individuals and their families relocated from urban to rural or semi-rural areas and these series of individual location-choice decisions prompted additional housing challenges, both in areas that lost residents and in the areas that gained them. The pandemic also prompted a re-thinking about the nature of paid work; as people worked at/from home, they discovered challenges ranging from the aesthetic to the practical need for auditory and visual privacy. They also sought safe recreational opportunities at or close to their place of residence. Yet, creating a backyard swimming pool or other kinds of play spaces for every single-family residential unit is neither feasible nor desirable from the perspective of climate change impacts – one of the three challenges identified at the beginning of this volume. In sum, housing policy professionals and housing advocates should be prepared for changes in the workplace that now include remote work and hybrid work arrangements that will directly impact housing preferences such as housing size and housing location and support services (Adikesavan and Ramasubramanian, in review, 2023).

Presently, housing activists are appropriately focused on tackling challenges of increasing residential density, which requires (i) building new housing, focusing on the alternatives to single-family houses on large lots, (ii) retrofitting existing housing to accommodate additional residential living units, (iii) changing zoning laws and local regulations to allow for different types of housing typologies in residential areas, and (iv) creating live-work spaces by blurring/weakening the rigid adherence to single-use zoning. However, housing activists and housing policy experts appear to disregard the cultural ethos that is deeply opposed to densification of the residential landscape. While we agree that all but the most rural counties in the United States should consider facilitating an increase in residential densities, we do not believe that

DOI: 10.1201/9780429243066-6

a singular focus on increasing residential density by outlawing single-family zoning will find favor with a majority of the American public.

Shaping public opinion in favor of higher densities requires housing policy professionals and urban planners to partner more closely with architects, landscape architects, and urban designers to create commodious interior spaces within a smaller footprint and create dense living environments that are visually appealing and are scaled to conform to the existing character of residential neighborhoods. We advocate for a deeper commitment to retrofitting and adapting the nation's older housing stock because the design of a typical single-family housing unit is not conducive for our nation's elderly to successfully age in place. Figure 6.1 (US Census, n.d.; NHGIS, n.d.) visualizes the spatial concentration of older homes, considering all residential living units, clustered in a large swatch of the Midwest and the Northeast. These areas are also highly correlated with the location of aging populations.

While the size of the total American population tripled since 1900, the population group aged over 65 years increased 11 times (Rivera-Hernandez et al., 2015). The retrofitting of single-family housing units to accommodate the day-to-day living needs of older adults is a national imperative, if we want to preserve the fabric of our neighborhoods. In other words, although the design/redesign/retrofit of the home may be perceived as the exclusive domain of architects and interior designers, the strong interconnectedness of housing with neighborhood level quality of life makes this issue relevant for housing reform advocates and urban planners. Research suggests that older people are happier living in their own home and that relocation to long term care facilities often reduces the quality of life (e.g., Cohen and Weisman, 1991; Wiles, et al., 2012). Yet older adults put off making much needed modifications and adaptations to making their homes safer for them (such as ramps, wider doorways, lower kitchen counters, etc.) because of social stigmatization around aging and being perceived as vulnerable (Bailey *et al.*, 2019).

In this context, housing policy analysts and housing advocates could encourage the retrofitting and ways to improve the quality of the housing within Naturally Occurring Retirement Communities (NORCs). NORCs have been identified as such since the 1970s (Hunt, 1998), first in New York and then with varying definitions in many states and at the federal level. Regardless of the specific numerical constraints (percentage of population, minimum number of seniors, and age threshold), the term "natural" is crucial because it (i) indicates that it is not a planned development (as in nursing homes, or purpose-built senior residence communities), (ii) the seniors involved have been living in those areas since before growing old, and (iii) in consequence of (i) and (ii) a NORC is ephemeral and will cease to be a NORC as its inhabitants cease to exist. NORCs can be located in aging suburbs, where individuals have aged in place, remaining in their homes after children left, or in newer suburbs as a result of migration among immigrant communities (especially Asian), where established and well-settled older children bring their parents from their home countries to live with them (Albrecht, 2007). The fleeting nature of NORCs can be problematic from a policy perspective because investments that attempt to address the special needs of such communities tend to experience a temporal lag. Bluntly put, during the time that elapses between the identification of a NORC, the allocation of special purpose funds (improving building infrastructure, specialized transportation services, etc.) and their implementation, many NORC residents may die without benefitting from such services, while the services/interventions will remain beyond the

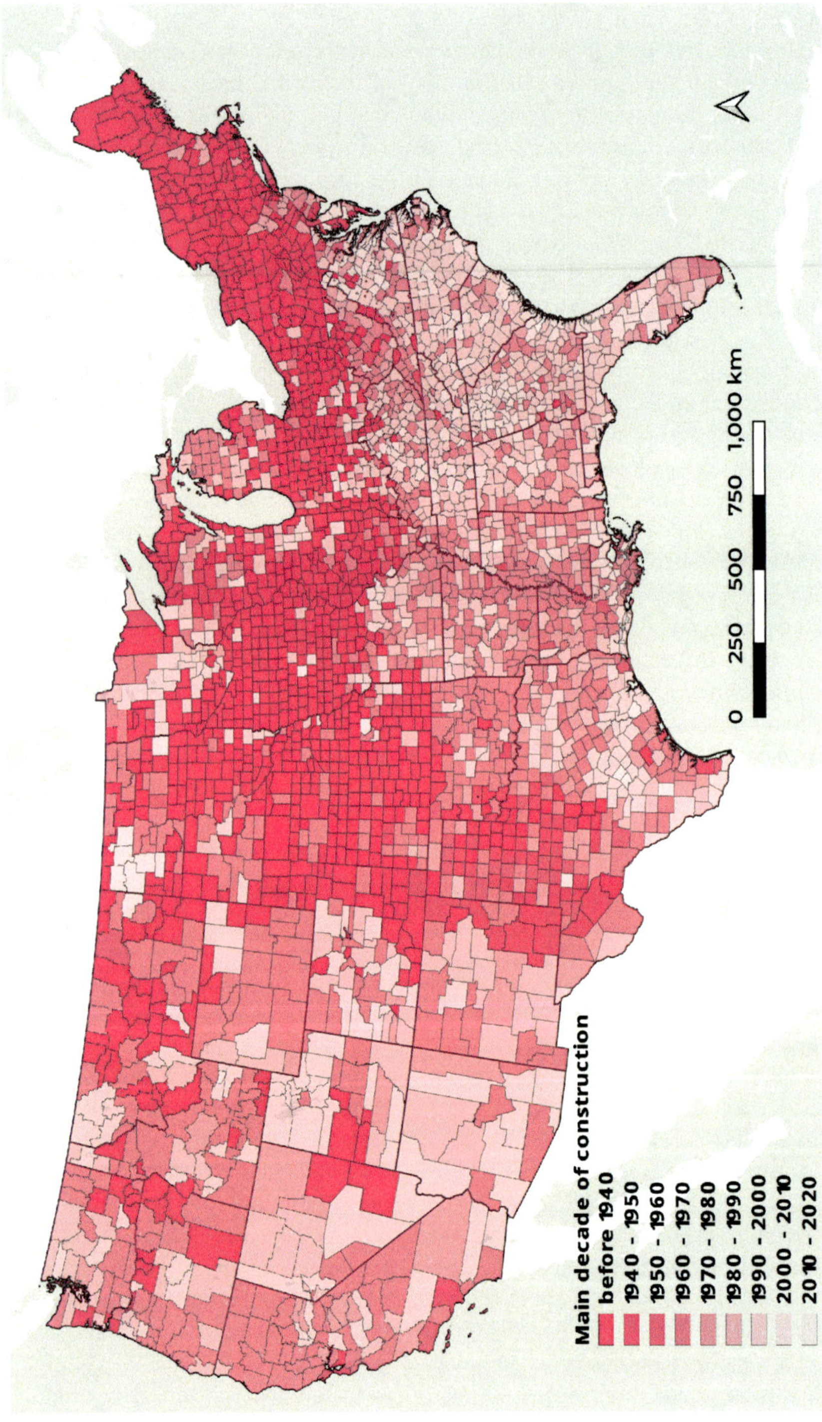


FIGURE 6.1 Building age – main decade when homes were constructed by county

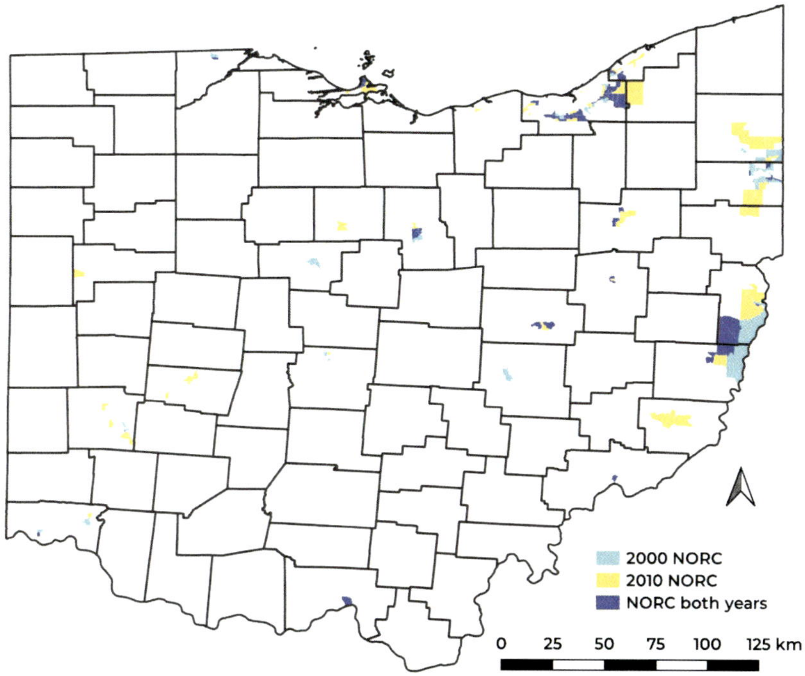


**FIGURE 6.2** NORC map of Ohio

cresting of the wave of local aging. However, considering that the American population is graying rapidly, investing in NORCs can encourage the creation of a great proportion of housing stock that includes the design and modifications to accommodate a wider range of aging adults by focusing on active aging (Scharlach, 2012).

The GIS implementation to identify NORCs is fairly straightforward. The US Census Bureau provides us with all the necessary data at the Census tract level. Depending on the age threshold chosen, we can calculate the percentage of those beyond that threshold as a percentage of the whole population of the respective tract. If that percentage is higher than the local or state regulations specify, we then have to check whether this is caused by the presence of nursing homes or other artificial distractors and subtract those residents from the calculated percentage. Alternatively, we could select the Census question "have you lived at this location ten years ago?" to determine whether the concentration is natural, see Figure 6.2 (ACS, 2010).

In terms of addressing the costs of owning and retrofitting single-family homes, architects and planners would do well to consider energy efficiency. The average age of a single-family home in New York is 60 years, while even in the state with the newest housing stock, Nevada, the average age is 23 years (NAHB, 2021), de facto assuring that the far majority of these homes are not particularly energy efficient. Depending on the materials used in the original construction, older homes cost more than modern homes to heat/cool. According to 2015 data provided by the *US Energy Information Administration*, single-family detached homes used 54% of their total energy consumption on space heating and air-conditioning while apartments with five or more units used only 32% for the same purposes. One way to increase energy efficiencies is for households to invest in

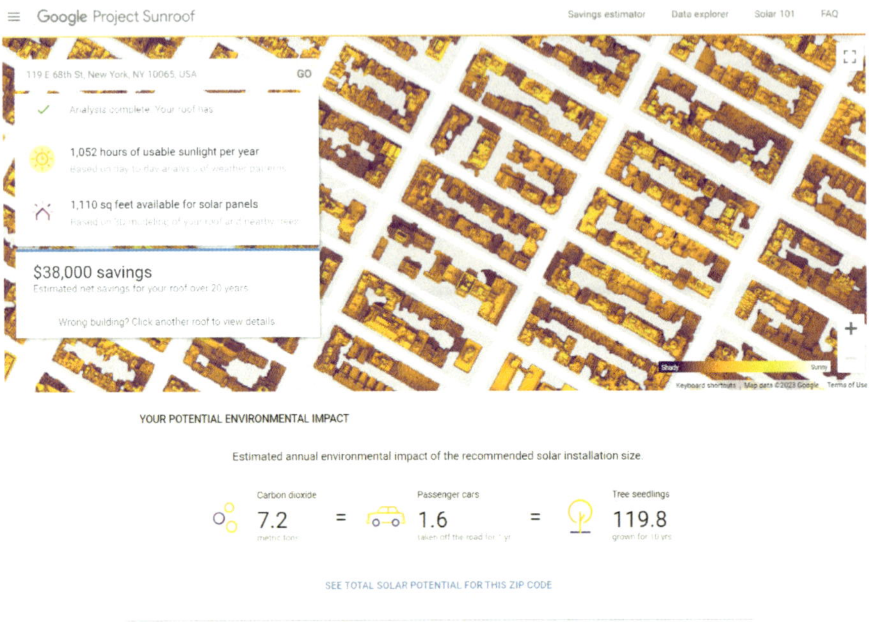


**FIGURE 6.3** Solar map of New York City

rooftop solar panels to reduce their draw on the power grid. GIS is an ideal instrument to calculate the solar potential of rooftop photovoltaic (PV) cells. The input data consists of aerial imagery and geophysical solar insolation measurements. The imagery is the basis for the calculation of the sizes and angles of roofs as well as the identification of any shading objects (trees and neighboring higher buildings). The combination of roof angle and insolation provides the amount of energy per area unit available. Given some stock measures of PV efficiency, the size of each roof, and the price for both the solar cells as well as local electricity rates, it is then straightforward to calculate the amortization time for each potential installation. Numerous states have released web maps that provide building owners with property-specific calculations (see, for example, https://nysolarmap.com/ or https://sunroof.withgoogle.com/), see Figure 6.3.

We have provided two examples where GIS can be used to strategically identify and improve the quality of life for individuals who live in single-family residential dwelling units. The next section will discuss potential GIS-based interventions at the neighborhood and community scale.

## 6.2 HOUSING AND NEIGHBORHOOD DESIGN

The single-family residential unit seldom exists as an island. Even the expansive and expensive estates of the wealthy are often integrated as part of cloistered enclaves. Housing scholars have observed that zoning regulations and restrictive covenants impact and influence these urban/suburban morphologies (Jackson, 1985). Housing policy specialists would benefit from acquiring a deeper understanding of these morphologies. Urban morphologies and subsequently suburban morphologies have been shaped by several factors, chief among them mobility patterns. Historically, as Southworth and Ben-Joseph (2003) write, two morphological patterns have dominated – the grid and the cul-de-sac, see Figure 1.13. The New Urbanism movement has long advocated a return to the grid, citing its superiority in improving walkability and reducing auto-dependence.

Most suburban layouts are set up as cul-de-sacs. While they can create a sense of safety for some, most suburban layouts that are designed with cul-de-sacs, coupled with the absence of sidewalks reduce mobility options for children, elderly, parents with small children, and anyone who may be temporarily or permanently disabled – forcing them to rely on an automobile to access the outside world. These communities emphasized the sanctity of the private sphere, privatizing available open spaces with visible fences or invisible boundary markers that cordon off space into developer-assigned lots. For groups of individuals or Home Owners Associations (HOAs) that want to create more opportunities for communal living and creating space-sharing arrangements, through a cohousing model discussed in Section 3.2.5, GIS can come in handy to facilitate the (i) identification of available shared spaces and (ii) reallocation and reestablishing of newer shared spaces to support play areas, kitchen gardens, and greenhouses. Specifically, the user would identify a study area, create a new layer of non-built-up space, and calculate the available acreage and create a centroid that equitably accommodates common/shared spaces. If HOAs are interested in adding sidewalks and bike paths to reduce auto-dependency, GIS tools can easily be deployed to identify optimal networks that connect individual properties without having to destroy existing built and green infrastructure.

The New Urbanists have consistently and with modest success sought to bridge the gap between design and policy, by demonstrating how design principles and building codes that are applied at the scale of a residential unit can be linked with neighborhood level regulations that can contribute to creating a neighborhood character without compromising individual autonomy. At the same time, they have advocated for including neighborhood codes that take into consideration building form, in addition to building use (Talen, 2011). Historically, new urbanism has focused on soft targets, implementing principles among people and communities that are already receptive to their ideas, such as those who are planning lifestyle or resort communities, or where the residents are affluent so as to not be concerned about the costs associated with emphasizing walkability, public transportation options, and creating neighborhood character. For older suburbs that are already walkable and built on a grid, GIS can be used as part of a neighborhood quality assessment toolkit (CNT, 2022). Urbanists and landscape architects such as Clare Cooper Marcus (1986) and Dolores Hayden (1980) have long argued that retrofitting suburbs is a viable social project that can yield many dividends. This is a societal project that cannot be successful in a top-down way, given how land use controls are managed at the local level. However, we are confident that easy-to-use GIS analyses and visualization tools can assist planners in small suburban communities to increase densities and improve accessibility options. For example, GIS software extensions such as *CommunityViz®* allow small towns and rural communities to have conversations about increasing density and assessing impacts on other variables associated with the quality of life (like traffic or school enrollment). It is much easier to conduct these types of analyses in small towns that operate as a distinctive local housing market. The Orton Family Foundation based in Vermont has developed a planning framework called Community Heart and Soul that relies on the use of GIS tools to translate values statements into assessment metrics.

However, successful urban design requires an assessment of space and place by considering massing, bulk, solids, and voids, essentially requiring a three-dimensional place analysis. Architectural software like SketchUp™ can create that immersive experience and these models can be incorporated as part of Google Earth visualization to situate a particular project in its real-world context. Such an approach is suitable for building scale projects, see Figure 6.4 (NYC Planning and NYCHA, 2020).

Although a true 3-d GIS is hard to come by, advanced parametric modeling approaches using software such as ArcGIS Urban or Rhino can create a neighborhood-level analysis of a cityscape. Regardless of their implementation, whether it is a building information model (BIM) or CityGML data, these implementations require a 3-d base map that in turn is generated from LiDAR data. This data generation process is reliant on outside experts as is working with (satellite) imagery data. We assume that the objects have been created to conform to existing data. Computer Generated Architecture (CGA) rules can create new analytically rigorous visualizations as depicted in Figures 6.5 and 6.6 (Kelly, 2021).

Working with building typologies as objects, we can now drop them onto a parcel or zoning map. Each object comes with a set of characteristics describing its services offered (floor space for specific functionality, energy efficiency, tax generated, etc.) as well as requirements (consumption values, demand on other services such as schools, hospitals, traffic, etc.). All of these characteristics then can be summarized by planning project indicators that describe the potential impact of a planning proposal (see Figure 6.7).

This is an advanced integration of GIS, 3-D modeling, and urban land use planning – while it can be used to evaluate policy proposals, there are very few American examples of this approach. The more innovative deployments of these techniques come from Asian countries like China and South Korea where city planners and architects use these approaches as part of their day-to-day work, see Figure 6.7.

## 6.3 EXPANDING INTRA- AND INTER-NEIGHBORHOOD MOBILITY ALTERNATIVES

The United States is a suburban nation (Bruegmann, 2005; Kruse and Sugrue, 2006) and the challenges of creating more housing have to directly engage with ways to densify suburbia. The term "suburbia" is widely used but poorly defined. There is an obvious relationship to 'urban' but sometimes it is part of the urban (vs. rural) fabric, while others see it juxtaposed to urban. Suburbia may be defined by its donut-like structure around the core of cities (in Europe often referred to as the "bacon belt"), with population densities and the subsequent provision of amenities that place it in the middle between fully urbanized and rural. A useful way to delineate suburbia is then to identify urban cores (places with >50,000 people and a population density of 7,500 people per square mile (appr. 3,000 people per km$^2$). We can then identify the surrounding areas where a threshold percentage of people commute to the urban core (the OECD, for instance, sets this threshold at 15%). Alternatively, if the commuting data is not available, travel time isochrones from the urban core may be used to delineate catchment areas. Using these measures, we arrive at the following Table 6.1 of suburban areas in the United States.

FIGURE 6.4 HUD defensible space

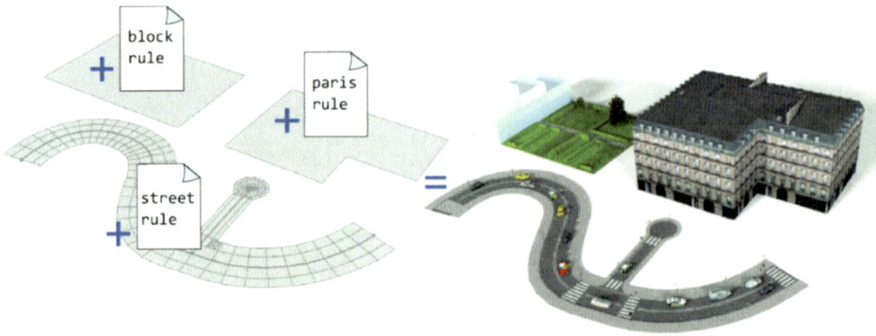


FIGURE 6.5 Combining CGA rules to develop a planning project

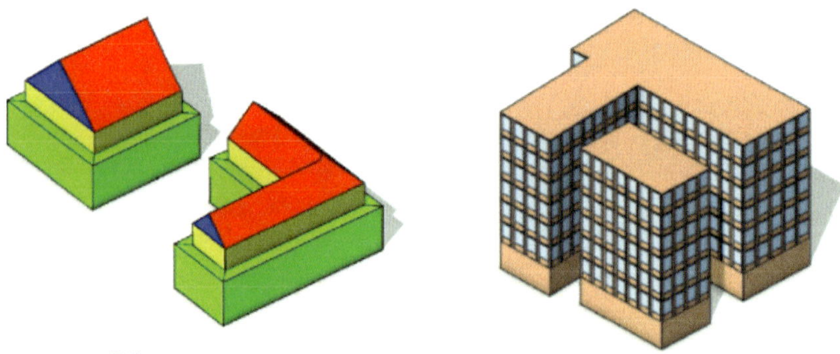

```
attr green  = "#00ff00"
attr yellow = "#ffff00"
attr blue   = "#0000ff"          # we'll also use blue. . .
attr red    = "#ff0000"          # . . . and red this time.

@Startrule
Lot  -->
  Lower

Lower  -->                       # rule 'Lower' :
   extrude (6)
   comp (f) {
      top : Upper |
      side : color(green) x.
   }
Upper  -->
   offset  (-1, 5)                # rule 'Upper' :
   comp(f)  (
      inside
         extrude (3)              # extrude a prise of height 3. . .
         comp (f) (               # separate te prise faces. . .
            top :
               Roof    |          # . . . the top face runs rule 'Roof' . . .
            side  :
              color(yellow)  x.   # . . .the sides remain yellow.
         }
      | border  :                 # as previously the border is green.
         color(green)  x.
   }

Roof  -->                         # rule 'Roof' :
   roofSable(45)                  # create a gabled roof.
   comp  (f)  {                   #separate the roof faces. . .
      vertical   :
        color (blue)  x.   |      # . . .vertical faces become blue. . .
      all  :
       color(red)  x.             # . . .and other faces become red.
   }
```

```
@Startrule
Lot -->                                   # rule 'Lot' :
     extrude (20)
     comp(f) {
          side : Facade |
          top  : Brick
     }

Brick -->                                 # rule 'Brick':
     color("#d29a78") x.

Facade -->                                # rule 'Facade':
      spit(y) {
          3.5: Brick |
          ~1 : Topfloors
      }
TopFloors -->                             # rule 'TopFloors':
      split(y) {
       ~3 : Floor
       }*

Floor -->                                 # rule 'Floor':
     split(x) {
        ~2: Tile                          # slices of the floor run Tile
     }*

Tile -->                                  # rule 'Tile':
   split (x) {                            # split tile sideways (x)
      0.3 : Brick |                       # left 30cm becomes brick
      ~1  : split(y) {                    # split the remainder verticlly
         0.8 : Brick |                    # bottom 80cm becomes brick
         ~1 : color ("#afc6e9")           # remainder is a blue 'window'
             x.   |
         0.2 : Brick                      # top 20cm becomes brick
      } |                                 # finish off the first (x) split
      0.3 : Brick                         # right 30cm becomes brick
   }
```

FIGURE 6.6 Examples for the parametric generation of housing objects based on CGA rules

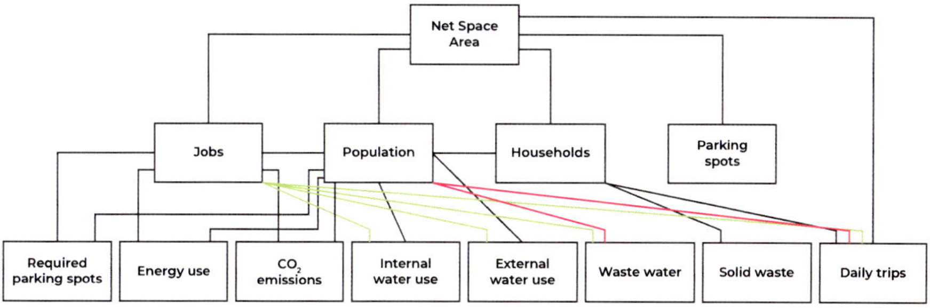


**FIGURE 6.7** Flowchart depicting the indicators required to determine the effects of a particular plan

**TABLE 6.1**
**The Size of Suburbia**

| 2020 Census Data | Area (km²) | Population |
|---|---|---|
| Urban core | 1,046 | 15,775,060 |
| Suburbia | 202,420 | 216,254,698 |
| Exurban/rural | 8,957,492 | 102,705,397 |

Yet, suburbia itself is not homogeneous and transportation choices shaped suburban development, see Figure 6.8 (SEDAC, 2019; MAP, 2015). The earliest suburbs, prominent in the densely populated north-east United States, were shaped by fixed transit lines that moved workers from outlying areas into Manhattan (New York) or Boston (Massachusetts). Walkability to and from the transit hub shaped these early suburbs. Automobile transportation allowed for further expansion and the cul-de-sac became a favored alternative.

Transforming car-dependent suburbs in the United States into walkable and bikeable neighborhoods requires a lot of planning and coordination between land use planning, housing, and transportation agencies, not to mention the commitment and involvement of the private sector (Dunham-Jones and Williamson, 2021). Such transformations will require physical changes like the introduction of sidewalks to improve walkability, and changes in local zoning laws to allow for mixed-use development, not to mention the provision for public transportation options. As we discussed in Section 5.5 of Chapter 5, cadastral data such as building age, planimetric data about the presence/absence of sidewalks resulting in the derivation of a walkability score, and zoning changes are starting points for a requirements analysis addressing future challenges in housing and neighborhood design. A full-fledged analysis would require a comprehensive agent-based modeling system (see Section 7.3 in Chapter 7).

While the re-design of suburban neighborhoods is a necessary step to alleviate the need for both new housing overall and different types of housing to serve diverse populations, movement within and between suburban neighborhoods deserves far more attention than it has in the past. Most suburban neighborhoods are entirely automobile dependent, and newer suburbs in most parts of the United States are sprawling

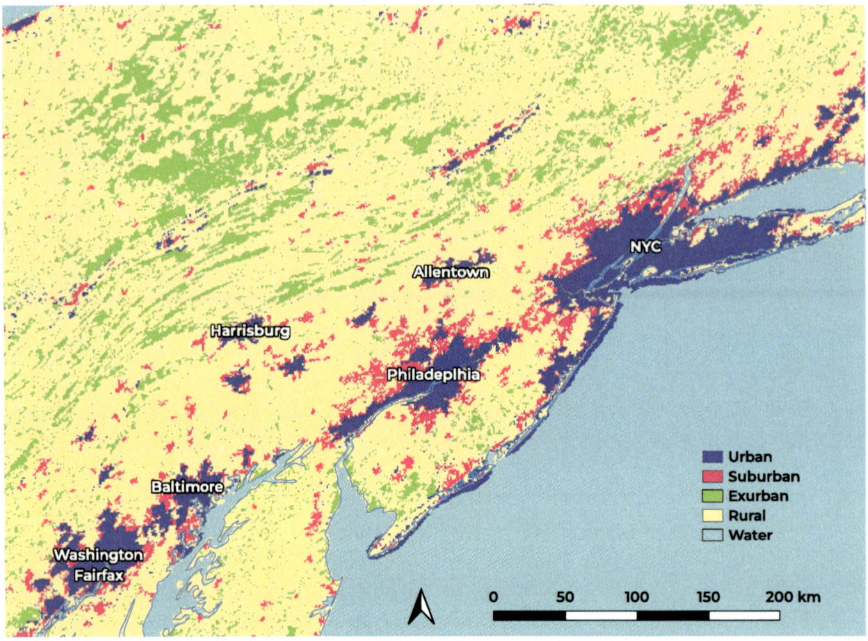


**FIGURE 6.8** Figure of Suburbia

sub-divisions that are not easily accessible on foot, even with the presence of pedestrian walkways. Furthermore, public transportation infrastructure in the United States was designed for a previous century where commuters were expected to travel (i) from the suburb to the city and back, (ii) leave and return at fixed times every weekday, and (iii) were going to the city to work in an "office". Our transportation systems have not easily adapted to the changing characteristics of commuting, and the changing nature of work. Fortunately, the newer solutions to solve this "last-mile problem" can be addressed by the provision of micro-mobility alternatives like e-bikes and scooters (at least for non-physical mobility-impaired populations). Now being pilot-tested in many areas around the country, this option is an affordable alternative to create intra- and inter-neighborhood mobility, reducing dependence on automobiles. In many parts of the country, it is not financially feasible to provide robust public transportation options; investments in just-in-time commute options can solve transportation problems for students, office workers, and low-wage workers who travel to suburbs to provide service work.

Additional complexities have been introduced as a result of post-pandemic shifts in the geography of US tech work. Jobs associated with the knowledge sector, and heavily aligned with Silicon Valley began to disperse as large firms like Alphabet (aka Google) and Meta (formerly Facebook) allowed their workers to work from anywhere. Workers dispersed to less expensive metropolitan areas in the Sunbelt and Mountain-West. Although this migration has great economic benefits to communities in Dallas, Denver, Orlando, Salt Lake City, Kansas City, St. Louis, and San Diego, it has created new ripple effects that housing policy professionals and housing advocates must seriously consider. Specifically, these mobilities are from higher density/higher cost metropolitan areas such as Los Angeles, San José and other Bay Area cities, Portland (Oregon), Seattle, or New York City to relatively low-density suburbs with lower housing costs, creating undesirable ripple effects such as reducing housing affordability and spurring gentrification (Adikesavan and Ramasubramanian, in review, n.d.; Florida and Kotkin, 2021; Muro and You, 2022; Peiser and Hugel, 2022). GIS tools can and have been successfully deployed to study the economic geographies of tech work (e.g., Zandiatashbar and Hamidi, 2022).

## 6.4 COMBATING HOMELESSNESS AND HOUSING INSECURITY

Homelessness in America is rising rapidly. A 2019 White House report stated that "over half a million people go homeless" on any single night in the United States. Although about 65% are temporarily housed in homeless shelters, the remaining 35% are living rough, on the streets. The vast majority of the homeless population, approximately 45%, is found in California, Hawaii, New York, Texas, Florida, Oregon, Washington, and the District of Columbia.[1] There is no "typical" homeless person. Men, women, children, elderly, people with disabilities, and veterans are part of the homeless population. The face of homelessness is the person on the street – typically an adult male, perhaps panhandling for change, talking to himself, or quietly suffering. Yet single adults actually constitute a minority of the city's homeless. The invisible face of homelessness is that of a child (ICPH, 2015). In New York City alone, approximately 28,000 school-age residents are living in shelters, 49,000 are living doubled up with other households, and 7,000 are living outside shelters or residences. These numbers derive from a survey of the NYC Dep of Education survey, which illustrates the degree to which official homelessness counts are underestimating the true dimensions of the homelessness problem. In addition to those who are actually without shelter, over 3.7 million people are experiencing housing insecurity, according to the National Alliance to End Homelessness, citing a Census Bureau survey (week 36, August 2021) that is tracking the impacts of the coronavirus pandemic. The shortage of affordable housing is an obvious cause – but the chronic homeless require more than shelter provisions – they need a bundle of services and support systems.

This suggests that the traditional way of counting the numbers and describing the problem is not helpful. GIS can be used both as a diagnostic as well as predictive tool, which in the hand of a GIS-savvy housing specialist provides the early warning signs that alert us to where interventions can be used to prevent homelessness. We suggest that *economic hardship, housing quality, stability*, and *affordability* are good indicators, which together provide a fairly accurate measure of where people live on the brink of homelessness. We discussed many of the necessary variables in Chapter 4, including the calculation of compound variables such as rent/mortgage burden, which may be countered by the availability of subsidized housing such as LIHTC. The lack of housing stability may be captured by any number of variables such as evictions, foreclosures, units whose rent subsidies expired, or just the percentage of new neighbors which can be derived from the US Census question, which is "How long have you lived at this address?". As Desmond (2017) describes vividly, homelessness is often precipitated by tenants living in places that eventually become unlivable. Crowding, building code violations, maintenance complaints, and increasingly common lists of bad landlords are excellent indicators of problems waiting to happen – especially to tenants whose landlords know that they don't have any other

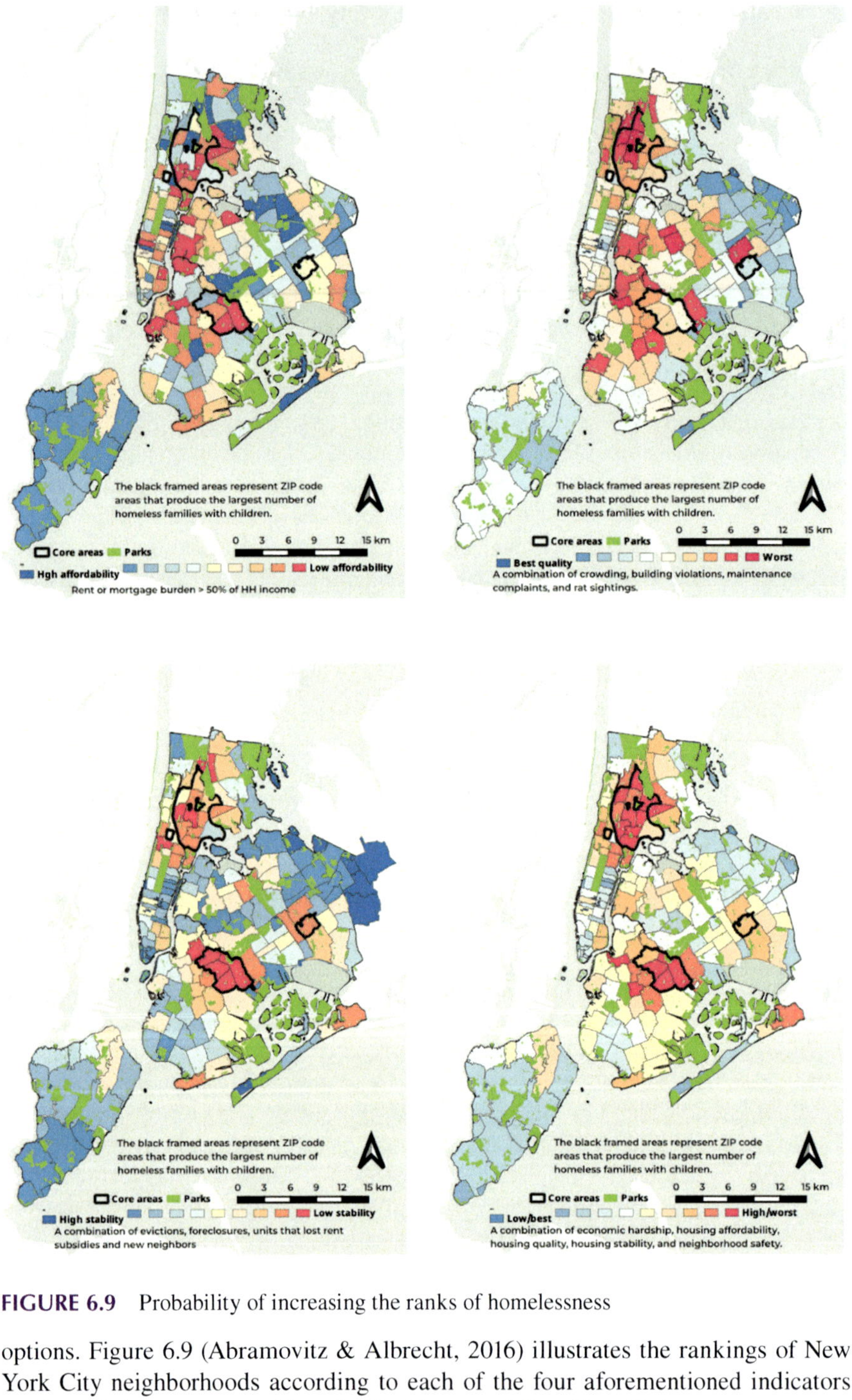


**FIGURE 6.9** Probability of increasing the ranks of homelessness

options. Figure 6.9 (Abramovitz & Albrecht, 2016) illustrates the rankings of New York City neighborhoods according to each of the four aforementioned indicators and how their aggregated effects in The Bronx and East New York.

New York City recently passed a local law that requires the local government to provide its citizens with information about a similar set of factors contributing to homelessness (or displacement risk as they put it). They aggregate four variables each to three higher-level categories: population vulnerabilities, housing conditions, and market pressures. The result can be explored online at https://equitableexplorer.planning.nyc.gov/map/drm/nta.

## 6.5 CLIMATE CHANGE AND SUSTAINABILITY

Global climate change has a significant impact on housing insecurity. In recent years, there have been dramatic disruptions in people's lives as weather patterns and climatic conditions have changed. There are climate-related catastrophes such as the destruction of entire neighborhoods and communities as a result of wildfires, flooding, and extreme heat. Climate change impacts and is impacted by the production of housing.

This last sentence deserves to be parsed carefully. One of the authors walked in October 2012 the promenade in Brighton Beach, NY, with their visiting relative, who asked how come there is no flood protection for the apartment high rises just across the street; one of the authors responded with "*because we have never experienced more than a foot of water*". A week later superstorm Sandy hit and hundreds of thousands had either fled or were trapped in their buildings. Unprecedented "natural catastrophes" are now occurring on an annual basis in one part of the country or another. Housing planners in a number of states are now busy developing buy-back plans to convince homeowners to move to less hazardous areas. And the storm-proofing of existing apartment complexes has become a new budget item that neither public nor private builders had never anticipated – not just in the Mississippi or Tennessee valleys but throughout the country. The question of where to allocate such resources is obviously a pertinent one. Yet, as we are looking to minimize the effects of climate change, we also need to be aware of the fact that housing itself is a driver of human-induced climate change. Urban sprawl contributes to climate change through higher emissions from land use change, embedded emissions in infrastructure, and transport energy consumption (NRDC, 2017; NREL, 2018; IPCC, 2021). Atmospheric $CO_2$ concentrations have reached a level that is unprecedented over the last 3 million years and the impacts of climate change are widely observed to be worsening globally (WMO et al., 2019). These impacts are strongly evident in cities, where urban policymakers and residents face extreme weather events – including heat waves, wildfires, flooding, and landslides – that particularly have an impact on vulnerable populations living in informal, low-quality, and overcrowded housing without the basic infrastructure, services, or green space that can offset the worst impacts of climate hazards (CUT, 2019). Those two aspects are intertwined when we look at the (need to) use air conditioning.

As we discussed in Chapter 2, air conditioning made large parts of the United States habitable. The building booms in the whole swath from Miami to Los Angeles would never have occurred without air conditioning. Yet, there are large parts of the country where people live without air conditioning, and this is about to change as climate change will alter the number of 100° days from a handful to several months. Figure 6.10 (First Street, 2022) depicts the counties where housing will have to adjust – preferably in such a way that it does not put an additional burden on an electricity grid that already struggles to provide charging stations for electric vehicles.

Organizations such as *First Street Foundation* are using GIS to perform parcel-level risk analyses for flood, fire, and heat hazards. While they are working with the Big Data

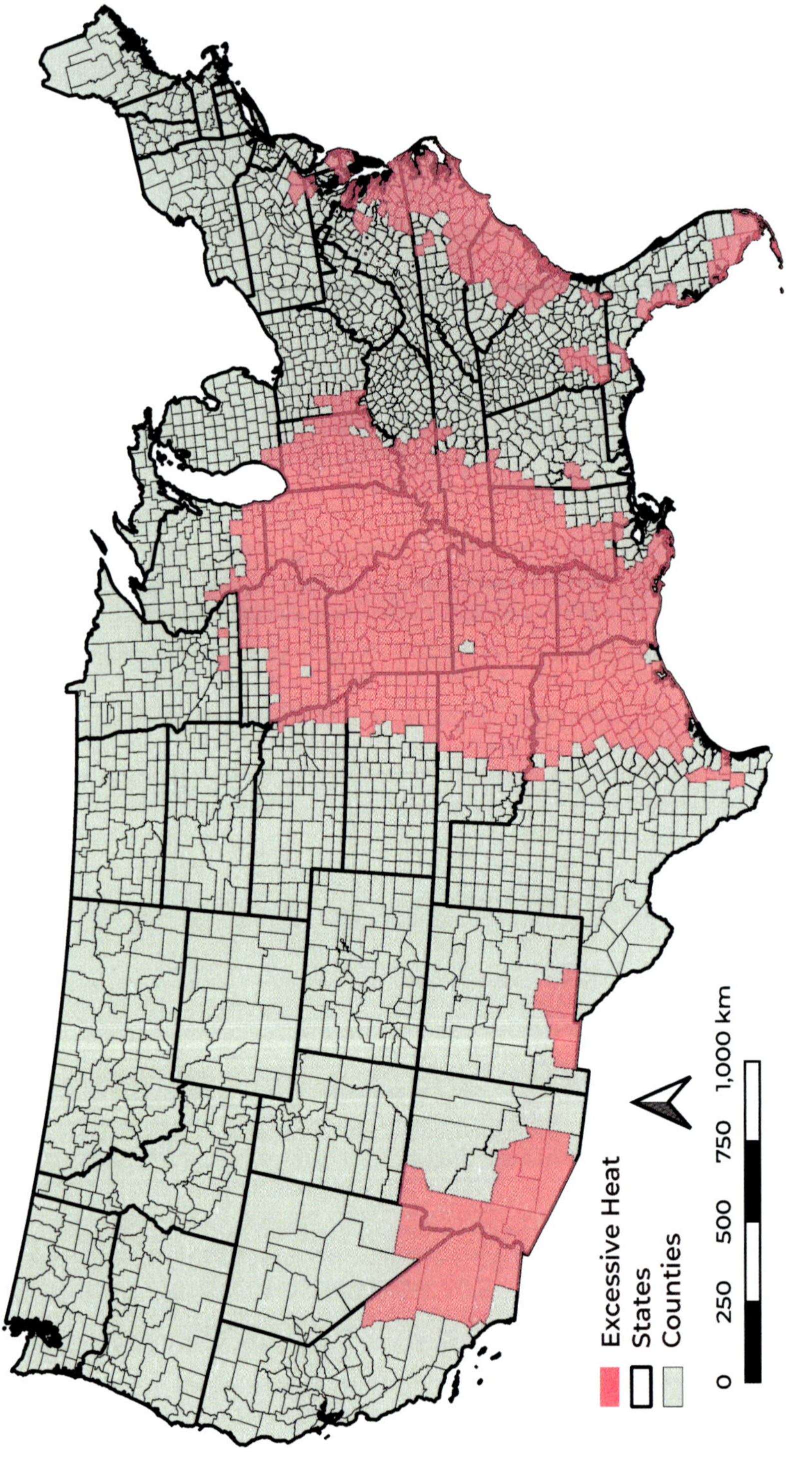


**FIGURE 6.10** Areas of excessive heat requiring adjustment in insulation and air conditioning

techniques further described in Section 6.7, local and regional planners have access to all the same public domain data and due to the smaller extent can then perform the same analyses and even improve on them because of their knowledge of local contexts. Quite similar to that private endeavor is the federal *Climate Mapping for Resilience and Adaptation* program, which is tract-level based but covers a wider range of calamities. As the flooding of Katrina or the heat waves in Chicago and Seattle have shown, climate change adaptation is a social equity issue. Intersecting (literally in the sense of a GIS operation as well as metaphorically as in addressing a housing issue from two different perspectives) such climate risk maps with social vulnerability factors will assist policy researchers with their prioritization in the allocation of sparse resources.

We mentioned the federal Hazus MH program before in Chapter 4 in the context of unusual data sources. The compilation of the data is an auxiliary function in service of its main purpose, which is hazard mitigation and management. Effective emergency response during or in the aftermath of a disaster is contingent on having a plan, which in turn requires having run scenarios of what is needed where in case of a disaster striking. We mentioned the surprise of superstorm Sandy before; the irony is that the NYC Department of Emergency Management had actually run a scenario of what would happen if the remnant of a hurricane is stalled by a blocking low-pressure system and that scenario predicted everything that was then actually happening. The scenario was considered too unlikely to invest the resources necessary to prevent the effects. But having run the scenario gave local and state authorities the information necessary to prioritize responses, which resulted in far fewer human casualties than Hurricane Katrina. GIS helps us to determine temporary shelter needs, even when local means of communication are interrupted because the geoprocessing models of systems like Hazus MH allow us to immediately calculate the follow-up effects of one resource outage or the other (e.g., gas station pumps not working when the electricity fails, preventing generators to be used as a substitute, or prioritizing the evacuation of mobility-impaired residents whose medical equipment at home is out of commission).

## 6.6 PUBLIC HEALTH AND SAFETY

The Covid pandemic provided us with examples for this delicate balance between public and private interests. Crowd control and enforcement of masking or vaccination requirements have first been modeled and then enforced using early adoptions of edge computing techniques. As in so many other spheres of life, the pandemic accelerated the adoption of techniques that otherwise would probably have taken decades to find acceptance. Two examples might illustrate this. In 2012, during Hurricane Sandy, some 80,000 residents of high-rise buildings, including elderly New Yorkers and those with physical limitations, were for 2 weeks stranded on upper floors when their buildings lost elevator service. Threats from water and food shortages, food poisoning from refrigeration not working, disease outbreaks from malfunctioning sewage systems/drinking water supply, and deficits in health care had become serious issues (Kunz et al., 2013) (see Figure 6.11 (Haraguchi & Kim, 2014)). And a repeat of the over 700 deaths during the 1995 Chicago heat wave (Klinenberg, 2002) is now unlikely even when we consider the climate change scenarios discussed in Section 6.5. The reason for that is that we now (potentially) have a much more detailed picture of vulnerable populations. The above experiences have led many local emergency response centers

**FIGURE 6.11** Map of elevator failures in public housing during superstorm Sandy

throughout the United States to continuously collect individual-level data of vulnerable populations – where vulnerability is a multidimensional measure based on age, race, health status, linguistic isolation, etc. The limitations these days are less a function of available technologies but of inter-departmental workflows that alert the appropriate administrative unit to potential dangers and trigger individualized responses.

This is a recurring theme: while GIS is an ideal medium to share data and trigger administrative actions, mental and procedural silos are limiting its use. Take the public health issue of walkability, for instance. Walkability maps are a type of map that shows the mobility of pedestrians in an environment. These maps can rate the walkability by different types of variables and generally include variables like proximity to amenities or public transit. Walkability maps can also be based on the characteristics of the physical environment, such as paved sidewalks or dangerous crossing locations, or on the convenience of walking to dining and drinking establishments, grocery stores, shopping, errands, parks, schools, and culture and entertainment.

This is a recurring theme: while GIS is an ideal medium to share data and trigger administrative actions, mental and procedural silos are limiting its use. Take the public health issue of walkability, for instance. Walkability maps are a type of map that shows the mobility of pedestrians in an environment. These maps can rate the walkability by different types of variables and generally include variables like proximity to amenities or public transit. Walkability maps can also be based on the characteristics of the physical environment, such as paved sidewalks or dangerous crossing locations, or on the convenience of walking to dining and drinking establishments, grocery stores, shopping, errands, parks, schools and culture and entertainment.

Albrecht et al. (2021) have shown that the Census Bureau's LODES data is representative not just for commuting but all kinds of trips. It can therefore be used to reflect the number, lengths, and modes of all forms of people's local and regional movements. In a separate study, Miller (2022) used the same LODES data to measure the effect of distance on movement mode. Looking at all commutes among the 51 neighborhoods of Brooklyn, NY, he found unsurprisingly a high correlation between the number of trips on foot and the density of residents and jobs in a neighborhood. More surprising is the amazing consistency of movement mode depicted in Figure 6.12: once locked into a mode, NYC commuters remain in that mode. At the same

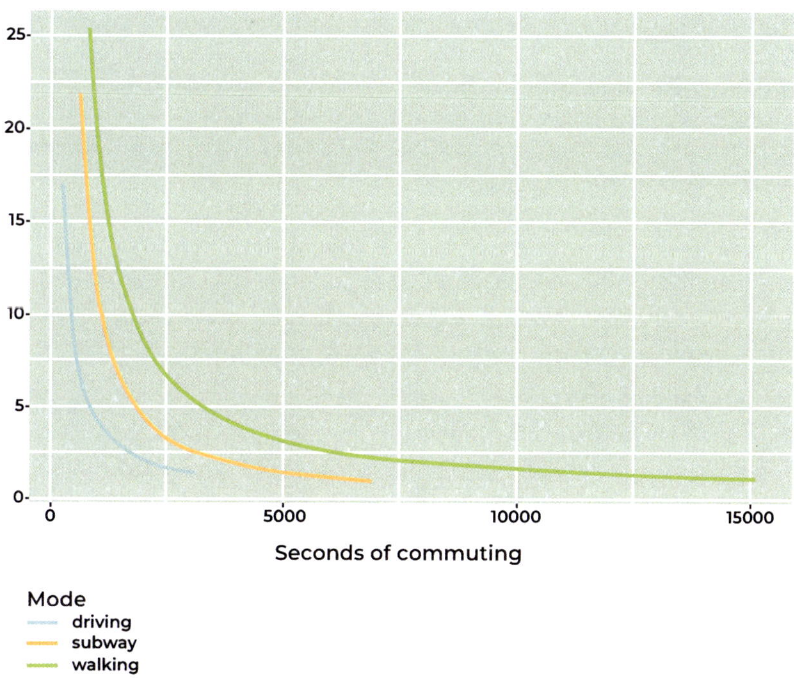


FIGURE 6.12 Movement mode as a function of commute density

time, Mr. Miller showed that compared to driving and walking, the use of public transportation showed the highest elasticity. Any decrease in commute time on the subway increases the number of commuters using that mode the most.

## 6.7 HOUSING AND SENSEABLE CITIES

Advances in AI and the ubiquity of sensor technologies help create a city that is a network of sensors that are passively "sensing" and gathering information about different aspects of the built environment. Just as a smart home can adjust its own temperature, turn lights off and on etc., both mobile and stationary devices can be deployed to improve the built environment. A variety of technologies come into play here: 5-G telecommunication, the so-called Internet-of-Things or IoT (expanded upon in the following), locational awareness, "Big Data" and associated methodologies such as neural networks and genetic algorithms combine to facilitate information flows without reliance on human intervention (see Figure 6.13 (EC, 2020)).

IoT architectures contain three layers:

- a perception layer consisting of sensors and actuators;
- a network layer that provides the communication between IoT devices and the Internet through Bluetooth or Wi-Fi; and
- an application layer either at the device level, within a local area network, or on some remote server.

From a housing policy perspective, the one aspect where GIS comes to bear is locational awareness. In the early 2000s, phone apps that allowed citizens to report outages were celebrated as a way to bridge the gap between citizens and the local authorities serving them. This is (or can) now be automated by way of sensors that report elevator outages or failing lights. Adopting the purpose behind a 311 call system, such

FIGURE 6.13 Schematic of information flows in a sensible city (based on European Commission, 2020)

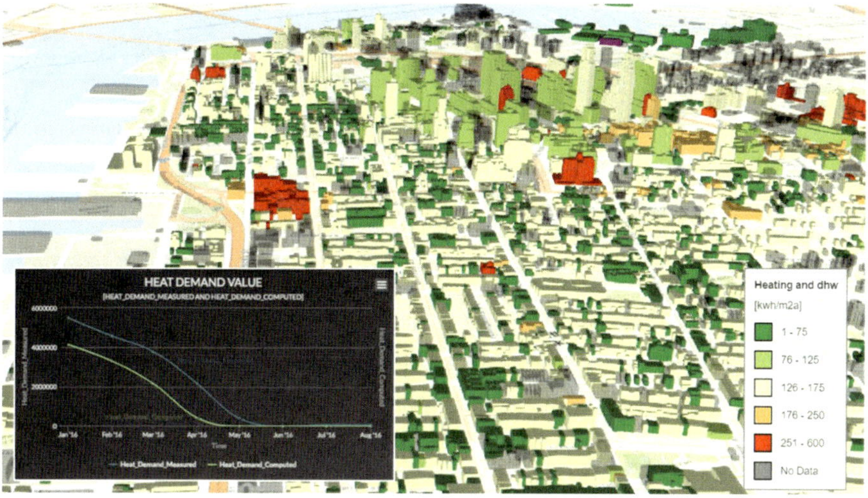


**FIGURE 6.14** Urban energy visualization of Brooklyn, NY

events can now be logged and service efficiency be measured. Especially with larger housing complexes or off-site landlords, such sensor-based maintenance and prevention promise a high return on investment. The combination of building information systems (BIM) and facility management results in what is now known as Smart Facilities Management (SFM), which has been successfully deployed in commercial office buildings (Gao and Pishdad-Bozorgi, 2019; Wang et al., 2022).

One aspect that Figure 6.14 above fails to reflect is the issue of scaling systems from the hyper-local (within a building) to the neighborhood or even city level. Each of the red keywords represents its own application development domain, which is typically unaware of the others. The link between SFM and city-wide models, say in the form of CityGML, has been implemented in a number of European pilot projects (as well as in China and Singapore) but is not a good fit in the US housing landscape. This leads to the ironic situation that European academicians such as Würstle et al., 2020) use open data repositories such as the one mandated in New York City as a rich source for their city-wide energy models.

Such models make use of the hierarchical organization of CityGML, which allows to scale information from individual windows and HVAC elements to nationwide building models such as Gilliland's 2019 Open City Model that covers every building in the United States. The energy model depicted in Figure 6.14 requires the same kind of information that we discussed in Section 6.2 Building and Neighborhood Design; the parameterization depicted in Figure 6.15 is akin to the CGA rules in Figure 6.5 but adjusted to the needs of an energy model.

Yet, the promise of Smart Cities remains so far largely unfulfilled. Most housing authorities were created many decades ago and are equipped with antiquated systems that are incapable of coping with the stream of data that sensors provide. The question now is who gets alerted, and do they have the means to react to the event triggered? An example of the need to adjust internal workflows to the changing I(o)T

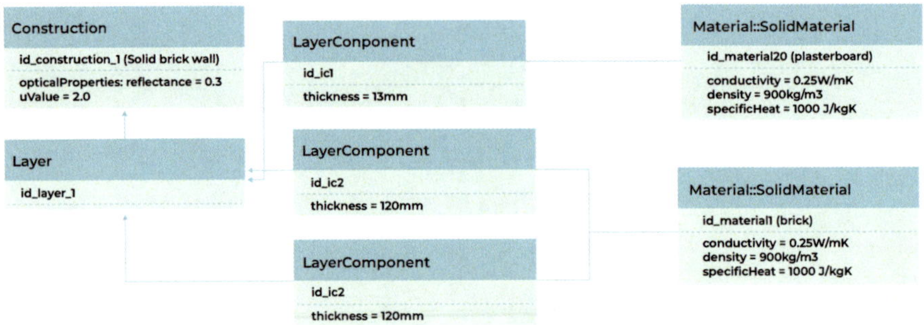


**FIGURE 6.15** Parameterization of a masonry wall as part of a CityGML based energy model

infrastructure is the New York City's Housing Authority's slow reaction to the detection of arsenic in drinking water (NY State Senate, 2022). In addition to the financial constraints of public authorities, there is the issue of perceived (and real) intrusion of privacy. One of the first applications of local sensors has been security cameras. Typically, in a public US context, they are used to record and act as a deterrent but in private as well as in East Asian environments, cameras are combined with face recognition to provide live access control. Western European and US authorities are, as of 2022, in the process of developing a regulatory framework to deploy federated or edge computing (Almutairi and Aldossary, 2021; Mondragón-Ruiz, Tenorio-Trigoso, Castillo-Cara, Caminero, and Carrión, 2021) that allows for decentralized local analysis and hence provide options to balance privacy with security options in local networks that do not require sharing with centralized servers.

The Scottish Cities Alliance defines a smart city as "the integration of data and digital technologies into a strategic approach to sustainability, citizen well-being and economic development" (Urban Foresight, 2016). Issues of cybersecurity, privacy, and sustainability, and public policy that prioritizes them, are central to understanding and successfully deploying smart city technology.

The vision of smart city services is built on data and system integration. These are the very same elements that make smart city infrastructures high-value targets for malicious actors. Their interdependent nature by design also means that attacks on one service frequently will have negative ripple effects on others. Incidents such as the Mirai botnet, which disabled a large part of the internet in 2016 have shown the vulnerability of multiple sensor networks to malicious interference (Wright, 2019).

Smart cities rely on using machine learning techniques; however, these techniques are prone to amplifying human biases that inform the design and training of such systems (Barocas and Selbst, 2016). Smart cities must function equally well for different stakeholder groups such as residents, commuters, and visitors to mention a few. City planners must ensure that bias in the machine learning ecosystem does not lead to systematically underserving identifiable sub-groups. Similarly, policy makers should target broad and fair access and application of machine learning techniques. This can be achieved through transparent planning and decision-making processes for smart city infrastructure and application developments, such as open hearings, focus groups, and advisory panels. The goal must be to minimize potential harm while maximizing the benefits that algorithmic decision-making can bring. The European Commission's High-Level Expert Group on Artificial Intelligence has published guidelines for trustworthy AI (EC, 2022) that addresses similar issues, including awareness of possible biases and harms and accountability.

Investment in smart cities has the potential to contribute significantly to achieving regional and global greenhouse gas emission reduction targets. Devices placed throughout cities, for example, can collect large volumes of data to enable coordinated decision-making toward more efficient use of resources (Barcelona, 2023). Ongoing research into both lower-energy devices and low-power wide-area networks to reduce the energy requirements of machine-to-machine communication shows promise but must be weighed against the energy needed to truly process Big Data. While the European Union, for instance, has committed to making all data centers net zero in carbon emissions by 2030 (EC, 2021), the path toward achieving this goal will be made more difficult by the growing amounts of data to store generated by and for smart cities.

## 6.8 CONCLUSION

In Chapter 6, we discussed new directions for future research that can leverage the spatial-analytical prowess of GIS to examine housing issues. Post World War II suburban morphologies have been created through an alliance of landowners, real estate developers, builders, car manufacturers, and politicians who believed that they were addressing the housing crisis of their time. We propose that GIS can be used by individual activists, nonprofit organizations, and housing policy professionals who want to explore various ways to transform and retrofit existing suburban neighborhoods to ameliorate and alleviate the problems of living in suburbia. GIS tools can be deployed to identify ways to make suburban environments walkable and bikeable, create safe routes to transit stops, or carve out shared open spaces. While beyond the scope of this book, we argue that GIS used alongside community organizing and mobilizing can be a powerful way to engage citizens in the physical transformations of residential environments in suburban contexts (Ramasubramanian, 2010).

GIS also provides housing policy experts and advocates alternative ways to engage the public on a series of housing related issues, by linking housing affordability and quality to public and environmental health, and to address the problems created by a changing climate. At the other end of the spectrum, digital twins allow planners and policymakers to model the impact of changing policies and changing physical interventions in real time within smart and senseable cities. While these innovations are still in a testbed phase, it's critical that we examine how access to data, data quality, and coherent communication pathways across disciplines are established. Housing is a key indicator in the evolution of smart city concepts with the idea that we advance 20-minute neighborhoods – neighborhoods that support a range of residential living choices, work opportunities, and recreation facilities, not to mention support services like educational and health care facilities. GIS anchors smart city modeling, especially as we strive to address societal concerns related to access and equity.

## NOTE

1. Some of these numbers are simply the result of the most populous states; others are a function of urbanization (homeless people tend to move from surrounding areas to urban centers), and climate (it is easier to survive in a non-freezing environment). Under-reporting is a function of acknowledgement/politics, i.e., the low numbers reported for Phoenix/Maricopa county don't withstand closer scrutiny.

## FURTHER READING

Albrecht, J, Petutschnig, A, Ramasubramanian, L, Resch, B, and Wright, A, 2021. *Comparing Twitter and LODES Data for Detecting Commuter Mobility Patterns.* MTI Technical Report 1946. San Jose, CA: Mineta Transportation Institute. Doi:10.31979/mti.2021.2037.

Broekhoff, D, Piggot, G, and Erickson, P, 2018. *Building Thriving, Low-Carbon Cities: An Overview of Policy Options for National Governments.* London and Washington, DC: Coalition for Urban Transitions. https://newclimateeconomy.net/content/cities-working-papers.

CUT, 2019. *Climate Emergency, Urban Opportunity: How National Government Can Secure Economic Prosperity and Avert Climate Catastrophe by Transforming Cities, Coalition for Urban Transitions (CUT).* London and Washington, DC: CUT. https://www.globalcovenantofmayors.org/wpcontent/uploads/2019/09/Climate-Emergency-Urban-Opportunity-report.pdf.

Rode, P, Heeckt, C, Ahrend, R, Huerta Melchor, O, Robert, A, Badstuber, N, Hoolachan, A, and Kwami, C, 2017. *Integrating National Policies to Deliver Compact, Connected Cities: An Overview of Transport and Housing. New Climate Economy Report.* https://www.coalitionforurbantransitions.org.

Rydge, J, Jacobs, M, and Granoff, I, 2015. *Ensuring New Infrastructure is Climate-Smart. Contributing paper for Seizing the Global Opportunity: Partnerships for Better Growth and a Better Climate.* London and Washington, DC: New Climate Economy. https://newclimateeconomy.report/misc/working-papers/.

WMO/GCP/UNEP/IPCC/ GFCS, 2019. *High-Level Synthesis Report of Latest Climate Science Information Convened by the Science Advisory Group of the UN Climate Action Summit 2019, World Meteorological Organization (WMO), Global Carbon Project (GCP), UN Environment Programme (UNEP), International Panel on Climate Change (IPCC), Global Framework of Climate Services (GFCS).* https://ane4bf-datap1.s3-eu-west-1.amazonaws.com/wmocms/s3fspublic/ckeditor/files/United_in_Science_ReportFINAL_0.pdf?XqiG0yszsU_sx2vOehOWpCOkm9RdC_ gN.

## REFERENCES

Abramovitz, M. and Albrecht, J. 2016. Place Matters – mapping shelter entries by New York City neighborhoods. Report prepared for NYC Center for Innovation through Data Intelligence, Office of the Deputy Mayor for Health and Human Services, NYC.

ACS, 2010. American Community Survey. Tract-level population count by age. Online resource available at https://census/gov/programs-surveys/acs.

Adikesavan, M. and Ramasubramanian, L., n.d. Planning for the Emerging Geography of Distributed Work. *Journal of Planning Literature* special issue on Planning in the Next Century. In review.

Albrecht, J, 2007a. *Key Concepts and Techniques in GIS.* London: Sage Publications.

Albrecht, J, 2007b. "The Changing Face of Naturally Occurring Retirement Communities". In: *Proceedings of the 2nd URISA GIS in Public Health Conference*, New Orleans, LA, 20–23 May, 2007.

Albrecht, J, Petutschnig, A, Ramasubramanian, L, Resch, B, and Wright, A, 2021. *Comparing Twitter and LODES Data for Detecting Commuter Mobility Patterns*. MTI Technical Report 1946. San Jose, CA: Mineta Transportation Institute. Doi:10.31979/mti.2021.2037.

Almutairi, J, and Aldossary, M, 2021. "A Novel Approach for IoT Tasks Offloading in Edge-Cloud Environments". *Journal of Cloud Computing*, 10: 28. Doi:10.1186/s13677-021-00243-9.

Bailey, C, Aitken, D, Wilson, D, Hodgson, P, and Douglas, B, 2019. What? That's for Old People, that Home Adaptations, Ageing and Stigmatisation: A Qualitative Inquiry." *International Journal of Environmental Research and Public Health*, 16 (24): 4989. Doi:10.3390/ijerph16244989

Barcelona, 2023. *Barcelona Digital City*. https://smartcity.bcn.cat/en/growsmarter.html, last accessed 30 May 2023.

Barocas, S, and Selbst, A, 2016. "Big Data's Disparate Impact." *California Law Review*, 104: 671.

Broekhoff, D, Piggot, G, and Erickson, P, 2018. *Building Thriving, Low-Carbon Cities: An Overview of Policy Options for National Governments*. London and Washington, DC: Coalition for Urban Transitions. https://newclimateeconomy.net/content/cities-working-papers, last accessed 27 May 2023.

Bruegmann, R, 2005. *Sprawl: A Compact History*. Chicago, IL: University of Chicago Press

Calzada, I, Pérez-Batlle, M, and Batlle-Montserrat, J, 2021. "People-Centered Smart Cities: An exploratory action research on the Cities' Coalition for Digital Rights". *Journal of Urban Affairs*, doi:10.1080/07352166.2021.1994861

CNT, 2022. *Geographic Information System (GIS) 101 Toolkit for Environmental Justice Organizations and Allies*. Chicago, IL: Center for Neighborhood Technology. https://cnt.org/publications/geographic-information-system-gis-101-toolkit-for-environmental-justice-organizations, last accessed 28 May 2023.

Cohen, U, and Weisman, G, 1991. *Holding Onto Home: Designing Environments for People with Dementia*. Baltimore, MD: Johns Hopkins Press

Cooper-Marcus, C, 1986. *Housing as if People Mattered: Site Design Guidelines for Medium-Density Family Housing*. Berkeley, CA: University of California Press.

CUT, 2019. *Climate Emergency, Urban Opportunity: How National Government Can Secure Economic Prosperity and Avert Climate Catastrophe by Transforming Cities*. London and Washington, DC: Coalition for Urban Transitions (CUT),. https://www.globalcovenantofmayors.org/wpcontent/uploads/2019/09/Climate-Emergency-Urban-Opportunity-report.pdf, last accessed 28 May 023.

Desmond, M, 2017. *Evicted: Poverty and Profit in the American City*. New York: Crown.

Dunham-Jones, E, and Williamson, J, 2021. *Case Studies in Retrofitting Suburbia: Urban Design Strategies for Urgent Challenges*. London: Wiley.

EC, 2020. *Europe Shaping the 5G Vision*. https://ati.ec.europa.eu/news/europe-shaping-5g-vision, last accessed December 2022.

EC, 2021. *'Fit for 55': Delivering the EU's 2030 Climate Target on the Way to Climate Neutrality*. https://eur-lex.europa.eu/legal-content/EN/TXT/PDF/?uri=CELEX:52021DC0550, last accessed 30 May 2023.

EC, 2022. *Ethics Guidelines for Trustworthy AI*. Brussels: Directorate General. https://ec.europa.eu/digital-single-market/en/news/ethics-guidelines-trustworthy-ai, last accessed 30 May 2023.

Eran, B, 1995. *Livability and Safety of Suburban Street Patterns: A Comparative Study*. Berkeley, CA: Institute of Urban and Regional Development, University of California, Working Paper 641.

First Street, 2022. https://firststreet.org/data-access/public-access/.

First Street Foundation, 2023. *Defining America's Climate Risk*. https://firststreet.org/, last accessed 31 May 2023.

Florida, R, and Kotkin, J, 2021. "America's Post-Pandemic Geography". *The City Journal*. https://www.city-journal.org/article/americas-post-pandemic-geography, last accessed 31 May 2023.

Gao, X, and Pishdad-Bozorgi, P, 2020. "BIM-Enabled Facilities Operation and Maintenance: A Review". *Advanced Engineering Informatics*, 39: 227–247. Doi:10.1016/j.aei.2019.01.005.

Gilliland, A, 2019. *Open City Model*. https://github.com/opencitymodel/opencitymodel, last accessed 12/24/2022 at

Hankin, C, 2022. *ACM TechBrief: Smart Cities*. New York: Association for Computing Machinery. Doi:10.1145/3534515.

Haraguchi, M. and Kim, S. 2014. *Critical infrastructure systems: a case study of the interconnectedness of risks posed by Hurricane Sandy for New York City*, prepared for the United Nations Office for Disaster Risk Reduction's Global Assessment Report on Disaster Risk Reduction 2015. Online resource available at https://www.preventionweb.net/publication/critical-infrastructure-systems-case-study-interconnectedness-risks-posed-hurricane.

Hayden, D, 1980. "What Would a Non-Sexist City Be Like? Speculations on Housing, Urban Design, and Human Work". *Signs* 5(3): S170–S187.

Hunt, M, 1998. "Naturally Occurring Retirement Communities". In: Shumsky, N (Ed.), *Encyclopedia of American Cities and Suburbs*, pp. 517–18. New York: Garland Publishing.

ICPH, 2015. *Beyond Housing: A National Conversation on Child Homelessness and Poverty*. https://www.icphusa.org/wp-content/uploads/2015/01/ICPH_UNCENSORED_6.1_Spring2015_ConferringonHomelessness.pdf, last accessed 21 May 2023.

Intergovernmental Panel on Climate Change (IPCC), 2021. *Climate Change 2021: The Physical Science Basis*. Contribution of Working Group I to the Sixth Assessment Report of the Intergovernmental Panel on Climate Change. Cambridge: Cambridge University Press.

Jackson, K, 1985. *Crabgrass Frontier: The Suburbanization of the United States*. London: Oxford University Press.

Kelly, T. 2021. City Engine: an introduction to rule-based modeling. In: Shi, W., Goodchild, M., Batty, M., Kwan, M. and Zhang, A. (Eds.) *Urban Informatics*. Singapore: Springer. doi:10.1007/978-981-15-8983-6_35.

Klinenberg, E, 2002. *Heat Wave: A Social Autopsy of Disaster in Chicago*. Chicago, IL: University of Chicago Press

Koebrich, S, Bowen, T, and Sharpe, A, 2018. *Renewable Energy Data Book*. U.S. Department of Energy (DOE), Office of Energy Efficiency & Renewable Energy (EERE). https://www.nrel.gov/docs/fy20osti/75284.pdf, last accessed 27 May 2023.

Kruse, K, and Sugrue, T, 2006. *The New Suburban History*. Chicago, IL: University of Chicago Press.

Kunz, M, Mühr, B, Kunz-Plapp, T, Daniell, J, Khazai, B, Wenzel, F, Vannieuwenhuyse, M, Comes, T, Elmer, F, Schröter, K, Fohringer, J, Münzberg, T, Lucas, C, and Zschau, J, 2013. "Investigation of Superstorm Sandy 2012 in a Multi-Disciplinary Approach". *Natural Hazards and Earth System Sciences*, 13, 2579–2598, doi:10.5194/nhess-13-2579-2013.

MAP, 2015. *Accessibility to Cities. Malaria Atlas Project* (University of Oxford). Online resource available at https://resourcewatch.org/data/explore/cit01701-Travel-Time-to-Major-Cities

Miller, T, 2022. *The Effect of Transportation Infrastructure on Trip Patterns throughout Brooklyn, NY*. Unpublished work conducted as part of a course on Spatial Analysis at Hunter College, City University of New York. https://tangoyankee.io/brooklyn-lodes/, last accessed 31 May 2023.

Mondragón-Ruiz, T-T, and Castillo-Cara, C, 2021. "An Analysis of Computational Resources of Event-Driven Streaming Data Flow for Internet of Things: A Case Study". *The Computer Journal*, 66(3): 47–60. doi:10.1093/comjnl/bxab143

Muro, M, and You, Y, 2022. *Superstars, Rising Stars, and the Rest: Pandemic Trends and Shifts in the Geography of Tech*. Washington DC: Brookings Institute. https://www.brookings.edu/research/superstars-rising-stars-and-the-rest-pandemic-trends-and-shifts-in-the-geography-of-tech/, last accessed 27 May 2023.

NAHB, 2021. *Median Age of Housing Stock by State Varies by More Than 35 Years*. National Association of Home Builders. https://www.nahb.org/blog/2021/04/median-age-of-housing-stock-by-state-varies-by-more-than-35-years/, last accessed 20 May 2023 at.

National Association of Home Builders (NAHB), 2022. *New Single-Family Home Size Continues to Grow*. https://www.nahb.org/blog/2022/03/new-single-family-home-size-continues-to-grow, last accessed 28 May 2023.

Natural Resources Defense Council (NRDC), 2017. Sprawl *Report 2017: Measuring Sprawl and Its Impact*. NRDC. Energy Data Book: Edition 37. U.S. Department of Energy.

New York State Senate, 2022. *Federal, State, and Citywide Elected Officials Write to Mayor and NYCHA Regarding Water at Jacob Riis Houses*. https://www.nysenate.gov/newsroom/articles/2022/brian-kavanagh/federal-state-and-citywide-elected-officials-write-mayor-an, last accessed 12/24/2022.

NHGIS, n.d. *Generic historical US Census data archived by National Historical GIS*. Online resource, available at nhgis.org.

NYC Planning and NYCHA, 2020. *Connected Communities Guidebook*. New York. https://on.nyc.gov/connected-communities

Peiser, R, and Hugel, M, 2022, "Is the Pandemic Causing a Return to Urban Sprawl?" *Journal of Comparative Urban Law and Policy*, 5(1): 26–41. https://readingroom.law.gsu.edu/jculp/vol5/iss1/7, last accessed 25 May 2023.

Ramasubramanian, L, 2010. *Geographic Information Science and Public Participation*. Heidelberg: Springer.

Ramasubramanian, L, and Albrecht, J, 2018. *Essential Methods for Planning Practitioners: Skills and Techniques for Data Analysis, Visualization, and Communication*. The Urban Book Series. Cham: Springer. doi:10.1007/978-3-319-68041-5

Rivera-Hernandez, M, Yamashita, T, and Kinney, J, 2015. "Identifying Naturally Occurring Retirement Communities: A Spatial Analysis". *The Journals of Gerontology. Series B, Psychological Sciences and Social Sciences*, 70(4): 619–27. doi:10.1093/geronb/gbu077.

Rode, P, Heeckt, C, Ahrend, R, Huerta Melchor, O, Robert, A, Badstuber, N, Hoolachan, A, and Kwami, C, 2017. *Integrating National Policies to Deliver Compact, Connected Cities: An Overview of Transport and Housing*. New Climate Economy Report. Coalition for Urban Transitions. https://newclimateeconomy.report/workingpapers/wp-content/uploads/sites/5/2017/12/NCE2017_OECD_LSE_NationalPolicies-1.pdf, last accessed 28 May 2023.

Rydge, J, Jacobs, M, and Granoff, I, 2015. *Ensuring New Infrastructure is Climate-Smart. Contributing paper for Seizing the Global Opportunity: Partnerships for Better Growth and a Better Climate*. London and Washington, DC: New Climate Economy. https://newclimateeconomy.report/misc/working-papers/, last accessed 26 May 2023.

Scharlach, A, 2012. "Creating Aging-Friendly Communities in the United States". *Ageing International*, 37: 25–38. doi:10.1007/s12126-011-9140-1.

SEDAC, 2019. *Disaggregated Population Density Data from the Socioeconomic Data and Applications Center at Columbia University (NY)*. Online resource available at https://sedac.ciesin.columbia.edu.

Southworth, M, and Ben-Joseph, E, 2003. *Streets and the Shaping of Towns and Cities*. Washington, DC. Island Press.

Talen, E, 2011. *City Rules: how Regulations Affect Urban Form*. Washington, DC: Island Press.

Urban Foresight, 2016. *Smart Cities Scotland Blueprint*. Newcastle: Urban Foresight. https://scottishcities.org.uk/wp-content/uploads/2021/01/Smart-Cities-Scotland-Blueprint.pdf, last accessed 29 May 2023.

US Census, n.d. *Generic American Community Survey (ACS)* data. Online resource, available at data.census.gov.

US Global Change Research Program (USGCRP), 2022. *Start Developing Your Climate Resilience Plan*. https://resilience.climate.gov/, last accessed 31 May 2023.

Wang, T, Gan, V, Hu, D, and Liu, H, 2022. "Digital Twin-Enabled Built Environment Sensing and Monitoring Through Semantic Enrichment of BIM with SensorML. *Automation in Construction*, 144: 104625. doi:10.1016/j.autcon.2022.104625.

Wiles, J, Leibing, A, Guberman, N, Reeve, J, and Allen, R, 2012. "The Meaning of 'Aging in Place' to Older People". *The Gerontologist*, 52(3): 357–366. doi:10.1093/geront/gnr098

WMO/GCP/UNEP/IPCC/ GFCS, 2019. *High-Level Synthesis Report of Latest Climate Science Information Convened by the Science Advisory Group of the UN Climate Action Summit 2019*. International Panel on Climate Change (IPCC), Global Framework of Climate Services (GFCS). https://ane4bf-datap1.s3-eu-west-1.amazonaws.com/wmocms/s3fs-public/ckeditor/files/United_in_Science_ReportFINAL_0.pdf?XqiG0yszsU_sx2vOe-hOWpCOkm9RdCgN, last accessed 27 May 2023.

Wright, R, 2019. *FBI: How We Stopped the Mirai Botnet Attacks*. Newton, MA: TechTarget. https://www.techtarget.com/searchsecurity/news/252459016/FBI-How-we-stopped-the-Mirai-botnet-attacks, last accessed on 29 May 2023.

Würstle, P, Santhanavanich, T, Padsala, R, and Coors, V, 2020. "The Conception of an Urban Energy Dashboard using 3D City Models". *e-Energy* 20, June 22–26, 2020, Virtual Event, Australia. doi:10.1145/3396851.3402650

Zandiatashbar, A, and Hamidi, S, 2022. "Exploring the Microgeography and Typology of U.S. High-Tech Clusters". *Cities*, 131: 103973. doi:10.1016/j.cities.2022.103973

# 7 Conclusions

## 7.1 INTRODUCTION

In the United States, academic conversations related to housing have appropriately focused on affordable housing and public housing – in other words – how to provide housing for those who are unable to successfully participate and thrive in the private property or rental markets (e.g., Schuetz, 2022) In large part, scholarly inquiry and public conversations about housing have challenged lawmakers to enact policies and programs that support the homeownership ideals while also mitigating risks (e.g., Belsky et.al, 2014). We encourage and support these conversations. Housing policy experts have examined the impacts of federal policies, and programs and many scholars want the federal government to be more involved in the provision of housing (e.g., Colburn and Aldern, 2022). Yet, the crisis remains.

Housing seems awash with data produced by academic think tanks, city agencies, nonprofit organizations, and advocacy groups. Yet, all these data don't appear to be producing new residential living units or preserving the housing stock we have. We came to the conclusion that academic housing experts have highly specialized and valuable knowledge about specific federal and state policies and programs; the historical and social context within which these housing policies are created, enacted, and sustained; and in-depth analyses of the impacts of housing policies on the lives of everyday people. It is these in-depth case studies and ethnographic narratives that reveal that it is very difficult to understand housing challenges, without simultaneously considering many other factors including demographics, health, education, and more specifically, *where* people live.

As authors who collectively have professional and practical expertise in the fields of architecture, urban planning, and Geographic Information Science (GIS), we are eager to directly assist those individuals, nonprofit groups, philanthropic organizations, and housing advocates who are doing the work of creating alternatives to avert the housing crisis – the acute shortage of housing alternatives that currently exists for all but the very wealthy. Chapter 1 articulates these challenges and provides a geographical framework to explain how everyone who is interested in solving housing challenges in their community can use GIS and spatial analysis to support and expand their sphere of influence. GIS allows end users to demystify housing policy and draw in those stakeholders whose engagement is sorely needed to create new housing alternatives. Chapters 2 and 3 speak to GIS professionals who are currently developing interesting analytical methods and using them to ask housing-related questions without having the historical context of how demographic change, urbanization, and federal policies and practices shaped the contemporary housing landscape.

The private housing and rental markets understand the power of location all too well. The infamous – location, location, location – mantra that every real estate broker makes when they present a property for sale short-circuits policy conversations and brings home the essential truth – the geographical (socio-spatial) context is critical

DOI: 10.1201/9780429243066-7

and implicit in any housing conversation. Yet, the geographical context that a realtor creates for the prospective buyer is a purpose-drawn "map" of an imaginary ideal; realtors typically tout the opportunities of any location under consideration, while downplaying constraints or limitations. Unlike a realtor's map, the GIS concepts we discuss in this book are rigorous and scientific. In Chapters 4 and 5 in particular, we provide you with a primer on how to build a GIS project from scratch – from data acquisition to the use of basic and advanced spatial analysis methods that encourage those who care about the quality of life in their cities and neighborhoods to participate thoughtfully in housing conversations.

Presently, housing policy experts tend to be isolated, often speaking within an echo chamber of like-minded people. In Chapter 6, we propose a different approach to drawing attention to our current crises of housing shortages and lack of affordability by encouraging research, analysis, and advocacy at the local level, where land use and zoning decisions are made. The reality is that contemporary housing policy is reactive and critical, more than it is visionary or even pragmatic. Since housing policy decisions reside within an interconnected framework of policy choices made by government and private entities, it may be prudent for housing specialists to engage and build alliances across disciplines and domains in creating new housing alternatives – whether it be accessory dwelling units, multi-family medium-rise housing units, co-housing models, or transfer of unused public land to create new affordable housing. GIS can support and facilitate the establishment of these connections.

We do not offer GIS as a panacea to address entrenched biases, including mistrust in government, racism, and prejudiced attitudes about who we want as our neighbors. We are also not favoring one set of housing policy or programmatic interventions over another. Our desire is to use GIS maps and associated analyses to communicate socio-spatial narratives to advance well-reasoned policy agendas. We are well aware that the United States in the 2020s is a hyper-polarized political landscape in which civil debate over ideas seems all but impossible. But we must try nonetheless. The alternatives are dire – rising housing costs, sprawl, environmental degradation, increased travel time, and overall reductions in the quality of life.

## 7.2 THE POWER OF GIS

Although GIS has been available as a set of tools since the 1970s, its initial promise to advance decision support was not fully realized until recently. We are now facing a split into two separate GIS communities: end users who have basic GIS functionality at their fingertips, be it in office software or web-based mapping on one side vs. power users, who mash up terabytes of data accessible through cloud services. GIS tools and functionalities have co-evolved alongside computational advances. We have been aiming for the middle. The power of GIS lies in its ability to combine data across departments and provenance. Some visualization packages such as Tableau have added an amazing array of mapping tools to their software. Linking disparate datasets is now often as easy as drag and drop. Yet, similar to the lament of statisticians, with powers come responsibilities. The distinction between basic and advanced GIS operations in Chapter 6 mirrors the separation between "democratized" GIS functionality now widely available and those operations that will remain the domain of GIS and related software (such as the R ecosystem). We might characterize the former as small letter gis to describe simple forms of communicating with maps and tell the difference from capital letter GIS, which requires conceptual models for the development of indicators and a keen eye on what is special about spatial analysis methods.

The conceptual models have to come from the application domain, whether this is housing proper or economic development, transportation, environmental protection, etc. We propose that housing policy analysts engage critically[1] with the use of GIS as they begin conversations about increasing density, for example, or about the siting of new affordable housing in residential neighborhoods. While there is a nascent Yes, in my backyard (YIMBY) movement emerging in areas of high unaffordability, the nation as a whole is largely resistant to high-density residential development.

None of the challenges listed in Chapter 6 can be solved without GIS. They require the collaboration between GIS novices, intermediate, and expert users with the housing policy analyst sitting in the middle. They need to be able to talk to Jane Public using storymaps, create her own analysis of policy interventions, and make conscientious use of datasets created by climatologists, epidemiologists, or economists. As useful as putting things on a map is from an exploratory perspective, the mere production of an atlas of all the different stakeholder perspectives (the "gis" from above) would be a severe short-selling of GIS's potential. GIS can be a communication platform that provides access to multiple expert knowledge bases and allows stakeholders to engage in constructive arguments about local decisions that in their multitude have regional impacts. It is the housing policy analyst's responsibility to use the advanced techniques discussed in 5.4 to put the relationships on a scientifically defensible quantitative basis. When stakeholders present their perspectives and lay open the data and methods used, decision making becomes transparent and the housing policy analyst fulfills her democratic mandate.

## 7.3 THE ROLE OF GEO-ARTIFICIAL INTELLIGENCE (GEO-AI)

We are taking a balanced view of the role of artificial intelligence (AI) in GIS in general and its application to housing research in particular. The term AI has been around for many decades and left a bad memory on most information scientists because of many broken promises and predictions that have proven to be wrong. The community has therefore been learning towards the adjective "computational" to connote the application of AI methods such as cellular automata, agent-based models, neural networks, or genetic algorithms in many disciplines. As of late 2022, so-called large language models have caught the attention of the general public, mostly by allowing them to retrieve facts and instructions in a conversational mode. A more technical audience has been using neural networks for object recognition in remotely sensed images or even video streams to update cadastral maps or provide live updates for crowd control.

"Segment Anything" (*Facebook 2023*), for instance, is the combination of a Python library and a carefully selected dataset that can be locally installed and for which there are numerous plugins to geospatial software packages (e.g., Wu and Osco 2023) that identify objects in images and depending on the tool that has been built on top of it, creates features in a variety of geospatial formats. Depending on the local hardware, this can be accomplished in real-time – a feat that eluded a generation of image processing and remote sensing researchers. The development of a customized model based on in-house imagery is probably beyond the brief of a housing researcher. But in collaboration with GIS staff, housing researchers can now develop their own customized deep learning models (MapFlow 2023[2]) based on multiple generations of imagery, possibly using a range of sensors (e.g., post-war black-and-white aerials, current drone photography, multi-spectral satellite imagery, and LiDAR) to perform analyses that would have sounded utopian a decade ago.

The GIS unit of the *Province of Cantabria* (Spain), for instance, has applied such object recognition techniques adapted to their own data holdings to create a storymap of *(sub-) urbanization*, to derive *3-D building objects*, or to perform *real-time crowd detection on their beaches* during the coronavirus pandemic.

In Section 6.2, we discussed the parametric generation of housing objects based on CGA rules. Podrasa *et al.* (2021) demonstrate how this can be scaled from individual objects (buildings) to the development of land use scenarios for generative urban design. The same way large language models such as GPT-4 work with so-called transformers that are capable of understanding the context of sequential data by analyzing the relationships between the words; their neural network uses generative approaches to implement what Cantrell and Mekies (2018) call "relational urbanism". Using building and neighborhood typologies parameterized as per our discussion in Section 6.2, their neural network worked through millions of possible combinations to create design solutions that outperformed every expert and Charette solution based on the criteria developed for a planning exercise in Berlin (Christ et al. 2017). Figure 7.1 illustrates the workflow of this generative approach.

## 7.4 EDUCATING THE NEXT GENERATION OF HOUSING ADVOCATES

Housing is an important area of specialization in graduate planning education in the United States. However, given the broad scope of topics covered within this subject area, students are unlikely to have a deep and immersive understanding of all aspects of the field unless they are getting a doctoral degree. Housing policy classes cover topics such as demographic trends, housing finance, public housing, fair housing, and community reinvestment, but they may not cover zoning for housing, design, and construction issues because those topics may be covered in a land use class or in an urban design class. While most graduate planning students in the United States now take at least one GIS class as they acquire their master's degree, their knowledge of "GIS" may be limited to basic mapping and analysis. This does not address the lack of GIS expertise among many of the housing planning or policy professionals and is exacerbated by the fact that many local authorities lack GIS experts in any department. Given the ubiquity of GIS functionalities built into housing related apps like Zillow, students may not even notice that they are being guided to explore housing problems in a market-driven way, rather than to consider variables that may result in a more equitable and community-oriented outcome.

21st century challenges like climate change or addressing income inequities require that policymakers use robust data to support integrative solutions.

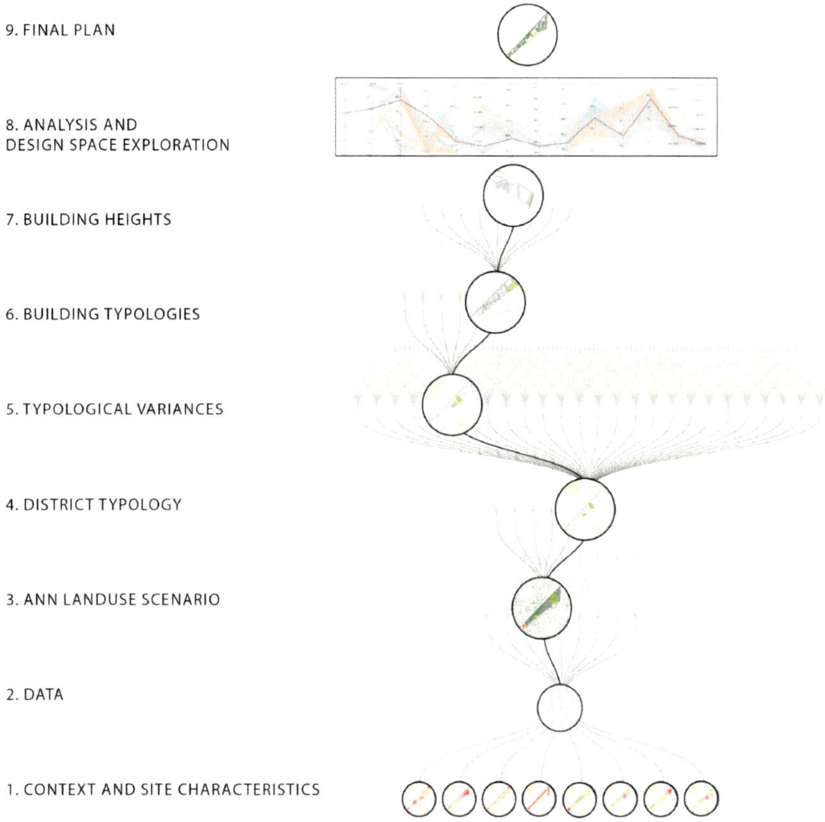


**FIGURE 7.1** Generative typology

For example, advocates of dense housing and advocates for public transportation can both come together to address climate change by advocating for dense housing to be built close to transit hubs. This thinking is not considered novel within academic planning circles – planning scholars have been advocating for such actions for many years. Yet, the case needs to be made to other groups of decision-makers, especially investors who are looking for solid returns on their investment. In addition to prospective home buyers and renters, engaging with the real estate industry about location decisions requires that housing policy professionals take a regional view, challenging the site-specific view that most real estate professionals use to push their projects through. GIS, as discussed in earlier chapters and Section 7.2, can support this shift in perspective.

## 7.5 WHERE TO FROM HERE?

We have to confront the sobering reality that there is a serious housing supply and affordability crisis in the United States, one of the wealthiest countries on the planet. The fragmented nature of land use planning and land management is one of the reasons that the United States struggles to produce a high volume of affordable housing. As we look at the socio-political landscape in 2023, it is hard to imagine a national consensus about tackling these housing crises. The most optimistic way forward is for a bottom-up approach of an intentional large-scale social transformation, anchored by creating walkable/bikeable and environmentally sustainable neighborhoods, cities, and states.

The pathway to creating meaningful changes to serve diverse housing needs and diverse populations requires that we do not adhere to a single ideology or a single planning strategy to move forward. We can continue to innovate by advancing the use of sustainable building materials, as well as promote the salvage and reuse of construction materials, and explore advances in environmentally friendly construction techniques. In the design sphere, creating housing using adaptive design and universal design principles to support healthy aging and aging in place is essential.

In policy terms, we must continue to have conversations about increasing residential building density, including a commitment to densifying suburbia. While it is challenging to undo the unsustainable suburban landscapes created in the 1950s to the 1990s, it is also critical. Densifying suburbia cannot simply focus on the housing infrastructure; by necessity, we have to also consider the capacity of street networks, water and sewer infrastructure, and services. Housing in already dense neighborhoods and communities requires that we address different policy challenges, including solving the burdens of housing affordability and consequent displacement and housing precarity. Recently, Democrats in California have proposed what is perceived as a bold move to advocate for a Viennese model of "social housing" we discussed in Section 2.13 where the city owns about 25% of the city's housing stock for low-income residents. Assembly Bill, AB 309 advocates for social housing to be used as a way to address the shortage of affordable homes for all income levels in California. There is a need to build more housing, build it quickly, and build it to accommodate low-income families so that they can live there if not in perpetuity, for a period of time to create stability for their families and allow for the creation of a sense of community. We are not opposed to the government getting involved in the housing construction and management business just as they were over a half century ago. Yet, we have to learn from the mistakes of the past to avoid repeating them.

GIS is often referred to as the science of "where" and in this book, we have described the power and promise of geographical analyses. We have taken a unique perspective and set of approaches to engage you, the reader, to explore how the use of Geographic Information Science concepts and methods can advance applied research and policymaking in housing. GIS tools can provide a bridge to establish connections between different fields and disciplines by connecting different conceptual frameworks using spatial anchor points that are familiar – building, neighborhood, city, and region to advance more equitable and just housing policies and practices.

## NOTES

1. We use the term here not in the sense of critical theory but in the sense of a spatially aware citizen who does not uncritically fall victim to the gospel of GIS vendors or blindly applies GIS functionalities because there is a button for that.
2. MapFlow and UrbanMapping projects of the GeoAlert company, online resource available at https://github.com/Geoalert, last accessed 28 May 2023.

## REFERENCES

Belsky, E, Herbert, C, and Molinksy, J, 2014. *Homeownership Built to Last: Balancing Access, Affordability, and Risk After the Housing Crisis*. Washington, DC: The Brookings Institution Press.

Cantrell, B, and Mekies, A, 2018. *Codify: Parametric and Computational Design in Landscape Architecture*. Milton, UK: Routledge.

Christ, W, Weihrauch, H, and Kahlert, B, 2017. *Der Urban INDEX Shopping: Urbane Mitte Pankow. Risiken und Potentiale einer handelsorientierten Stadtentwicklung*. Darmstadt, Germany: Urban INDEX Institut.

Colburn, G, and Aldern, C, 2022. *Homelessness Is a Housing Problem: How Structural Factors Explain U.S Patterns*. Oakland, CA: University of California Press.

Facebook, 2023. *Segment Anything*. Facebook.

Github, 2023. *MapFlow and UrbanMapping projects of the GeoAlert Company*. https://github.com/Geoalert, last accessed 28 May 2023

Podrasa, D, Zeile, P, and Neppl, M, 2021. "Machine Learning for Land Use Scenarios and Urban Design". In *CITIES 20.50—Creating Habitats for the 3rd Millennium: Smart—Sustainable—Climate Neutral. Proceedings of REAL CORP 2021, 26th International Conference on Urban Development, Regional Planning and Information Society,* pp. 489–498. REAL CORP.

Schuetz, J, 2022. *Fixer-Upper: How to Repair America's Broken Housing Systems*. Washington DC: Brookings Institution Press.

Wu, Q, and Osco, L, 2023. *Samgeo: A Python Package for Segmenting Geospatial Data With the Segment Anything Model (SAM)*. Zenodo. doi:10.5281/zenodo.7966658.

# Appendix 1

**TABLE A.1**
**Housing Tenure Variables of the American Community Survey**

| Table | Title |
|---|---|
| **B07013** | Geographical Mobility in the Past Year by Tenure for Current Residence in the U.S. |
| **B07413** | Geographical Mobility in the Past Year by Tenure for Residence 1 Year Ago in the U.S. |
| **B08137** | Means of Transportation to Work by Tenure |
| **B08537** | Means of Transportation to Work by Tenure |
| **B17019** | Poverty Status of Families by Household Type by Tenure |
| **B25003** | Tenure |
| **B25007** | Tenure by Age of Householder |
| **B25008** | Total Population in Occupied Housing Units by Tenure |
| **B25009** | Tenure by Household Size |
| **B25010** | Average Household Size of Occupied Housing Units by Tenure |
| **B25011** | Tenure by Household Type (Including Living Alone) and Age of Householder |
| **B25012** | Tenure by Families and Presence of Own Children |
| **B25013** | Tenure by Educational Attainment of Householder |
| **B25014** | Tenure by Occupants Per Room |
| **B25015** | Tenure by Age of Householder by Occupants Per Room |
| **B25016** | Tenure by Plumbing Facilities by Occupants Per Room |
| **B25020** | Tenure by Rooms |
| **B25021** | Median Number of Rooms by Tenure |
| **B25022** | Aggregate Number of Rooms by Tenure |
| **B25026** | Total Population in Occupied Housing Units by Tenure by Year Householder Moved Into Unit |
| **B25032** | Tenure by Units in Structure |
| **B25033** | Total Population in Occupied Housing Units by Tenure by Units in Structure |
| **B25036** | Tenure by Year Structure Built |
| **B25037** | Median Year Structure Built by Tenure |
| **B25038** | Tenure by Year Householder Moved Into Unit |
| **B25039** | Median Year Householder Moved Into Unit by Tenure |
| **B25042** | Tenure by Bedrooms |
| **B25043** | Tenure by Telephone Service Available by Age of Householder |
| **B25044** | Tenure by Vehicles Available |
| **B25045‡** | Tenure by Vehicles Available by Age of Householder |
| **B25046** | Aggregate Number of Vehicles Available by Tenure |
| **B25049** | Tenure by Plumbing Facilities |
| **B25053** | Tenure by Kitchen Facilities |
| **B25106** | Tenure by Housing Costs as a Percentage of Household Income |
| **B25115** | Tenure by Household Type and Presence and Age of Own Children |
| **B25116** | Tenure by Household Size by Age of Householder |

*(Continued)*

**TABLE A.1 (*Continued*)**

**Housing Tenure Variables of the American Community Survey**

| Table | Title |
|---|---|
| **B25117** | Tenure by House Heating Fuel |
| **B25118** | Tenure by Household Income |
| **B25119** | Median Household Income the Past 12 Months by Tenure |
| **B25120** | Aggregate Household Income by Tenure and Mortgage Status |
| **B25123** | Tenure by Selected Physical and Financial Conditions |
| **B25124** | Tenure by Household Size by Units in Structure |
| **B25125** | Tenure by Age of Householder by Units in Structure |
| **B25126** | Tenure by Age of Householder by Year Structure Built |
| **B25127** | Tenure by Year Structure Built by Units in Structure |
| **B25128** | Tenure by Age of Householder by Year Householder Moved Into Unit |
| **B25129** | Tenure by Year Householder Moved Into Unit by Units in Structure |

**TABLE A.2**

**Housing Value Variables of the American Community Service**

| Table | Title |
|---|---|
| **B25075** | Value |
| **B25076** | Lower Value Quartile (Dollars) |
| **B25077** | Median Value (Dollars) |
| **B25078** | Upper Value Quartile (Dollars) |
| **B25079** | Aggregate Value (Dollars) by Age of Householder |
| **B25080** | Aggregate Value (Dollars) by Units in Structure |
| **B25082** | Aggregate Value (Dollars) by Mortgage Status |
| **B25083** | Median Value (Dollars) for Mobile Homes |
| **B25096** | Mortgage Status by Value |
| **B25097** | Mortgage Status by Median Value (Dollars) |
| **B25100** | Mortgage Status by Ratio of Value to Household Income |
| **B25107** | Median Value by Year Structure Built |
| **B25108** | Aggregate Value (Dollars) by Year Structure Built |
| **B25109** | Median Value by Year Householder Moved Into Unit |
| **B25110** | Aggregate Value (Dollars) by Year Householder Moved Into Unit |
| **B25121** | Household Income by Value |

## TABLE A.3
### Mortgage-Related Variables of the American Community Survey

| Table | Title |
|---|---|
| **B25027** | Mortgage Status by Age of Householder |
| **B25081** | Mortgage Status |
| **B25082** | Aggregate Value (Dollars) by Mortgage Status |
| **B25087** | Mortgage Status and Selected Monthly Owner Costs |
| **B25088** | Median Selected Monthly Owner Costs (Dollars) by Mortgage Status |
| **B25089** | Aggregate Selected Monthly Owner Costs (Dollars) by Mortgage Status |
| **B25090** | Mortgage Status by Aggregate Real Estate Taxes Paid (Dollars) |
| **B25091** | Mortgage Status by Selected Monthly Owner Costs as a Percentage of Household Income |
| **B25096** | Mortgage Status by Value |
| **B25097** | Mortgage Status by Median Value (Dollars) |
| **B25098** | Mortgage Status by Household Income |
| **B25099** | Mortgage Status by Median Household Income |
| **B25100** | Mortgage Status by Ratio of Value to Household Income |
| **B25101** | Mortgage Status by Monthly Housing Costs as a Percentage of Household Income |
| **B25102** | Mortgage Status by Real Estate Taxes Paid |
| **B25103** | Mortgage Status by Median Real Estate Taxes Paid (Dollars) |
| **B25120** | Aggregate Household Income by Tenure and Mortgage Status |

## TABLE A.4
### Variables Related to Home Ownership Costs in the American Community Survey

| Table | Title |
|---|---|
| **B25087** | Mortgage Status and Selected Monthly Owner Costs |
| **B25088** | Median Selected Monthly Owner Costs (Dollars) by Mortgage Status |
| **B25089** | Aggregate Selected Monthly Owner Costs (Dollars) by Mortgage Status |
| **B25091** | Mortgage Status by Selected Monthly Owner Costs as a Percentage of Household Income |
| **B25092** | Median Selected Monthly Owner Costs as a Percentage of Household Income |
| **B25093** | Age of Householder by Selected Monthly Owner Costs as a Percentage of Household Income |
| **B25094** | Selected Monthly Owner Costs |
| **B25095** | Household Income by Selected Monthly Owner Costs as a Percentage of Household Income |

## TABLE A.5
### Variables Related to Rent Costs in the American Community Survey

| Table | Title |
|---|---|
| **B25031** | Median Gross Rent by Bedrooms |
| **B25057** | Lower Contract Rent Quartile (Dollars) |
| **B25058** | Median Contract Rent (Dollars) |
| **B25059** | Upper Contract Rent Quartile (Dollars) |
| **B25060** | Aggregate Contract Rent (Dollars) |
| **B25062** | Aggregate Rent Asked (Dollars) |
| **B25064** | Median Gross Rent (Dollars) |
| **B25065** | Aggregate Gross Rent (Dollars) |
| **B25066** | Aggregate Gross Rent (Dollars) by Units in Structure |
| **B25067** | Aggregate Gross Rent (Dollars) by Meals Included in Rent |
| **B25070** | Gross Rent as a Percentage of Household Income |
| **B25071** | Median Gross Rent as a Percentage of Household Income (Dollars) |
| **B25072** | Age of Householder by Gross Rent as a Percentage of Household Income |
| **B25074** | Household Income by Gross Rent as a Percentage of Household Income |
| **B25111** | Median Gross Rent by Year Structure Built |
| **B25112** | Aggregate Gross Rent (Dollars) by Year Structure Built |
| **B25113** | Median Gross Rent by Year Householder Moved Into Unit |
| **B25114** | Aggregate Gross Rent (Dollars) by Year Householder Moved Into Unit |

## TABLE A.6
### Variables Describing Housing Characteristics in the American Community Survey

| Table | Title |
|---|---|
| **B25031** | Median Gross Rent by Bedrooms |
| **B25041** | Bedrooms |
| **B25042** | Tenure by Bedrooms |
| **B25068** | Bedrooms by Gross Rent |
| **B25016** | Tenure by Plumbing Facilities by Occupants Per Room |
| **B25047** | Plumbing Facilities for All Housing Units |
| **B25048** | Plumbing Facilities for Occupied Housing Units |
| **B25049** | Tenure by Plumbing Facilities |
| **B25050** | Plumbing Facilities by Occupants Per Room by Year Structure Built |
| **B25051** | Kitchen Facilities for All Housing Units |
| **B25052** | Kitchen Facilities for Occupied Housing Units |
| **B25053** | Tenure by Kitchen Facilities |
| **B25054** | Kitchen Facilities by Meals Included in Rent |

**TABLE A.7**

**Esri's Tapestry Segmentation Categorized by Racial/Ethnic and Housing Characteristics and Their Geographic Distribution**

| Market Segment | Dominant Geography | Renting % | Households Total | White % | Asian % | Black % | Hispanic % |
|---|---|---|---|---|---|---|---|
| High-rise renters | Poor, multi-generational dense cities | 96.3 | 622,500 | 23.8 | 4.9 | 36.5 | 57.4 |
| Fresh Ambitions | Multi-generational immigrants in older major city neighborhoods | 72.9 | 794,600 | 35.2 | 5.7 | 24.4 | 55.2 |
| Farm to Table | Hispanic agricultural, mostly in CA and WA | 55.6 | 299,600 | 44.6 | 2.6 | 3.4 | 82.9 |
| Family Extensions | Older Hispanic neighborhoods Periphery of West Coast metros, NYC and CHI | 64.3 | 912,400 | 43.7 | 4.2 | 6.3 | 84.7 |
| NeWest Residents | Recent immigrants in large metros in South and West | 83.4 | 970,800 | 44.4 | 4.5 | 11.3 | 72.2 |
| City Commons | Low-income mid-sized buildings in metro cities | 77.0 | 1,106,600 | 14.3 | 1.3 | 75.9 | 9.1 |
| Southwestern Families | Hispanic older neighborhoods in SW city centers and suburbs | 46.3 | 1,021,400 | 69.8 | 82.5 | 5.9 | 1.6 |
| Forging Opportunity | Urban periphery of larger metros in South and West | 40.5 | 1,289,900 | 56.2 | 2.4 | 7.9 | 72.5 |
| Modest Income Houses | Older urban neighborhoods in the eastern half of the country | 55.3 | 1,627,600 | 10.1 | 0.5 | 84.7 | 4.8 |
| Hometown Heritage | Old neighborhoods in central cities in South and Midwest | 60.0 | 1,507,700 | 53.2 | 2.3 | 28.0 | 20.8 |
| Social Security Set | Older housing in high-density metro cities | 86.2 | 1,001,400 | 49.7 | 7.1 | 31.2 | 18.6 |
| Diverse Convergence | Dense urban peripheries on the Coasts and CHI | 72.4 | 1,528,100 | 44.0 | 11.3 | 11.7 | 57.6 |
| Metro Fusion | Urban periphery apartments | 76.0 | 1,753,500 | 42.7 | 5.2 | 30.8 | 34.2 |
| City Strivers | Dense city neighborhoods Bos-Wash and Chi | 68.1 | 962,900 | 12.2 | 3.0 | 71.3 | 19.4 |
| Down the Road | Semi-rural mobile homes in metro areas in South and West | 34.8 | 1,406,700 | 67.7 | 1.8 | 10.7 | 27.5 |
| Downtown Melting Pot | High-density apartments in CA and Mid-Atlantic | 68.6 | 814,000 | 43.0 | 39.1 | 4.3 | 20.1 |
| Rural Bypasses | Ultra-rural South | 30.1 | 1,646,400 | 56.5 | 0.5 | 34.7 | 5.6 |

(*Continued*)

**TABLE A.7 (*Continued*)**

**Esri's Tapestry Segmentation Categorized by Racial/Ethnic and Housing Characteristics and Their Geographic Distribution**

| Market Segment | Dominant Geography | Renting % | Households Total | White % | Asian % | Black % | Hispanic % |
|---|---|---|---|---|---|---|---|
| Urban Villages | Older homes in urban periphery of larger metros | 19.9 | 1,319,200 | 49.5 | 11.2 | 7.4 | 63.4 |
| Family Foundations | Stable low-income neighborhoods in cities S and W | 34.3 | 1,299,600 | 13.2 | 1.1 | 79.8 | 7.4 |
| Urban Edge Families | Urban periphery of larger metros in South and West | 36.3 | 1,824,900 | 52.1 | 5.4 | 20.0 | 44.4 |
| Small Town Sincerity | Small towns and semi-rural neighborhoods | 50.3 | 2,305,700 | 76.5 | 1.3 | 13.2 | 10.4 |
| Front Porches | Old neighborhoods | 53.5 | 1,960,300 | 63.7 | 4.8 | 15.0 | 24.0 |
| Set to Impress | Suburban apartments | 72.3 | 1,714,100 | 64.7 | 3.7 | 18.5 | 16.7 |
| Traditional Living | Low-density urban areas in Midwest and South | 41.1 | 2,395,200 | 74.3 | 1.7 | 13.4 | 12.7 |
| Dorms to Diplomas | Older small apartment buildings | 92.5 | 630,300 | 70.6 | 12.9 | 9.8 | 8.5 |
| Economic Bedrock | Ultra-rural mining and mobile homes | 24.5 | 810,000 | 84.2 | 0.6 | 6.3 | 11.7 |
| Senior Escapes | Seasonal rural homes in CA AZ, and FL | 24.8 | 1,116,000 | 93.9 | 1.6 | 4.4 | 13.9 |
| College Towns | Dense student housing in mid-sized cities and towns | 75.4 | 1,176,200 | 71.6 | 7.8 | 12.4 | 10.2 |
| Young and Restless | Dense city neighborhoods in non-coastal areas | 86.9 | 2,131,500 | 53.3 | 8.0 | 23.9 | 22.5 |
| Southern Satellites | Rural exclaves in Southern metros | 22.3 | 3,856,800 | 84.1 | 0.8 | 7.9 | 8.8 |
| Rooted Rural | Rural Appalachia, TX and AR | 20.2 | 2,430,900 | 88.4 | 0.5 | 5.8 | 5.1 |
| Heartland Communities | Rural areas from Rustbelt to Great Plains | 30.6 | 2,850,600 | 88.4 | 0.9 | 4.7 | 6.8 |
| City Lights | Dense urban but not apartment | 48.3 | 1,813,400 | 60.2 | 13.5 | 10.6 | 25.7 |
| Old and Newcomers | Gentrifying city neighborhoods | 54.8 | 2,859,200 | 76.5 | 3.9 | 10.9 | 11.9 |
| Retirement Communities | No particular geography | 54.9 | 1,501,100 | 79.2 | 4.6 | 9.2 | 11.6 |

(*Continued*)

**TABLE A.7 *(Continued)***

**Esri's Tapestry Segmentation Categorized by Racial/Ethnic and Housing Characteristics and Their Geographic Distribution**

| Market Segment | Dominant Geography | Renting % | Households Total | White % | Asian % | Black % | Hispanic % |
|---|---|---|---|---|---|---|---|
| Trendsetters | Cores of high-rent cities | 75.5 | 1,319,400 | 57.7 | 14.8 | 11.4 | 24.3 |
| Bright Young Professionals | Apartments in urban periphery of larger metros | 57.2 | 2,750,200 | 65.1 | 6.4 | 16.6 | 17.4 |
| Rustbelt Traditions | Dense fringe of metros in South and Midwest | 28.8 | 2,716,800 | 81.2 | 2.1 | 8.8 | 11.5 |
| Pacific Heights | Urban periphery of CA and Northeast metros | 27.6 | 889,400 | 34.7 | 48.6 | 3.1 | 15.6 |
| Parks and Rec | Older suburban neighborhoods | 30.3 | 2,449,600 | 78.7 | 3.7 | 0.6 | 12.3 |
| Middleburg | Semi-rural places within metros | 26.6 | 3,511,200 | 79.5 | 2.4 | 10.0 | 11.2 |
| The Great Outdoors | Rural areas in West, South and Northeast | 22.5 | 1,908,600 | 87.4 | 1.7 | 3.0 | 8.7 |
| Home Improvement | Low-density suburbs | 20.6 | 2,114,500 | 69.3 | 5.7 | 13.8 | 19.7 |
| Up and Coming Families | New suburban peripheries | 26.1 | 2,901,200 | 63.7 | 6.9 | 15.3 | 27.3 |
| Midlife Constants | Older suburban periphery of small metros | 27.3 | 3,068,400 | 86.0 | 2.1 | 6.6 | 7.7 |
| Salt of the Earth | Rural areas in OH, PA, IN | 16.9 | 3,545,800 | 93.0 | 0.7 | 2.6 | 3.8 |
| Rural Resort Dwellers | Scenic rural, often seasonal | 18.9 | 1,227,200 | 92.0 | 0.8 | 2.1 | 5.1 |
| Prairie Living | Ultra-rural in the Midwest | 20.7 | 1,323,200 | 92.8 | 0.6 | 1.1 | 6.6 |
| Pleasantville | Suburbs of larger coastal metros | 16.9 | 2,718,100 | 73.1 | 8.5 | 8.8 | 17.6 |
| Enterprising Professionals | Suburbs all over | 48.8 | 1,737,200 | 54.1 | 23.3 | 12.1 | 14.7 |
| Emerald City | Low-density neighborhoods in all urban areas | 51.5 | 1,748,600 | 77.7 | 5.2 | 9.3 | 11.1 |
| The Elders | Suburban periphery of warm metros | 18.6 | 910,100 | 93.1 | 1.8 | 2.7 | 5.6 |
| Military Proximity | Metro suburbs South and West | 97.0 | 186,600 | 65.3 | 4.6 | 16.9 | 18.5 |
| Metro Renters | Urban cores | 79.8 | 1,911,500 | 66.9 | 14.5 | 10.8 | 11.7 |

*(Continued)*

**TABLE A.7 (*Continued*)**

**Esri's Tapestry Segmentation Categorized by Racial/Ethnic and Housing Characteristics and Their Geographic Distribution**

| Market Segment | Dominant Geography | Renting % | Households Total | White % | Asian % | Black % | Hispanic % |
|---|---|---|---|---|---|---|---|
| Golden Years | Large metro areas but not central cities | 37.3 | 1,657,400 | 81.3 | 7.0 | 6.6 | 8.8 |
| Green Acres | Rural areas within metros | 13.9 | 3,923,400 | 90.8 | 1.6 | 3.3 | 5.5 |
| In Style | Older neighborhoods in metro cities | 32.2 | 2,764,500 | 83.5 | 4.6 | 6.1 | 7.8 |
| Comfortable Empty Nesters | Suburbs and small towns | 13.1 | 3,024,200 | 87.2 | 2.7 | 5.8 | 6.6 |
| Workday Drive | Suburban peripheries | 15.1 | 3,541,300 | 78.0 | 6.2 | 8.7 | 11.7 |
| Silver and Gold | Seasonal suburban near metro cities | 16.8 | 942,900 | 92.3 | 2.2 | 2.3 | 5.8 |
| Urban Chic | Suburbs of larger coastal metros | 33.8 | 1,635,200 | 79.1 | 9.7 | 4.3 | 10.2 |
| Boomburbs | Suburbs of larger metros | 16.0 | 2,004,400 | 68.1 | 15.6 | 8.0 | 15.0 |
| Laptops and Lattes | Cities in larger metro areas | 62.7 | 1,307,500 | 76.1 | 12.8 | 4.7 | 9.3 |
| Exurbanites | Suburbs of larger metros | 15.1 | 2,398,200 | 86.3 | 5.7 | 3.3 | 7.4 |
| Savvy Suburbanites | Suburbs of larger metros | 9.4 | 3,664,200 | 85.5 | 6.0 | 0.3 | 7.2 |
| Professional Pride | Suburbs of larger metros | 8.4 | 1,982,300 | 78.7 | 12.6 | 4.3 | 6.9 |
| Top tier | Suburbs of larger coastal metros | 9.8 | 2,113,000 | 82.8 | 11.2 | 2.2 | 5.9 |

# Index

**Note: Bold** page numbers refer to tables; *italic* page numbers refer to figures.

Made in the USA
Las Vegas, NV
17 September 2026

f5c700db-b95c-4b57-a3bc-4036772ba043R04